EXPOSÉ ON SUSTAINABILITY

I0601629

James H. Speer

Sustainability Press

TERRE HAUTE, INDIANA

 Sustainability
Press

Terre Haute, IN 47802

Publisher's Note: This is a work of fiction. Names, characters, places, and incidents are a product of the author's imagination. Locales and public names are sometimes used for atmospheric purposes. Any resemblance to actual people, living or dead, or to businesses, companies, events, institutions, or locales is completely coincidental. The characters, dialog, and actions in this book are made up to tell an interesting story, but the science and citations are accurate throughout. The real names of the scientists are used when talking about their work and extensive references are provided to the scientific articles as well as video clips demonstrating the issues that are discussed. The conversations are made up and do not claim to represent the opinions or actual conversations with these scientists.

Book Layout © 2019 BookDesignTemplates.com

Exposé on Sustainability/James H. Speer -- 1st ed.
ISBN 978-1-7336648-0-6

Dedication

I dedicate this book to my family. To my mother who taught me to care for the Earth. To my wife whom I love dearly and to my children who will hopefully benefit from a society that adopts a sustainable mindset.

CONTENTS

CHAPTER ONE

The Demise of Small Towns

Robertsville, Indiana
Saturday in April

I'm Greg Cunningham: a successful freelance writer who published in a variety of magazines and newspapers, but mainly submitted work in the New York Times. I was remembering the high of completing my last big exposé, which resulted in toppling one of the world's largest oil companies [1]. That story broke open the vested interests behind climate change and actually changed the CEO of the company from an adversary to an ally who continued to improve corporate operations for the benefit of the climate and human society. Extreme Oil Company abandoned their old ways and their dogged clinging to the outdated technology of petroleum. They moved full force into carbon sequestration, disbursed solar and wind power, and challenged Tesla with competing technologies in new battery development.

Jenny and I had moved out of downtown Indianapolis into a small town to the northwest of the city. This was the quaint old town of Robertsville that still had many buildings from the 1800s. We liked living here because it had its own grocery store, hardware store, and all of the amenities that you really needed. If you wanted to shop for something special or find good ethnic food, we would drive back into the city.

We were out for a Saturday drive on a warm spring day in our Prius. We needed a few items from the grocery store so we headed up to our favorite local shop called Corner Grocer. We knew the owners, Nancy and George, from our frequent visits. Shopping here was a personal experience and it felt like family in their shop. You could browse and get what you need, but you could also hang out and catch up on the local gossip. Nancy and George knew all of the comings and goings in the area.

"That doesn't seem right, Jenny. Look, the lights are off," I said.

"Let's go look in the window. Maybe they left a sign," she said.

We walked up to the front of the store. The town was eerily quiet for a Saturday morning on a nice spring day. We found a sign on the window saying "Going out of Business Sale".

"What!?! They can't be going out of business. They've always been so vibrant and a lot of people come into the store when we're visiting," Jenny said.

Just then, Nancy and George drove up in their old truck that they used for deliveries.

"Hey Nancy, George. What's going on? What is this going out of business sign?" I asked.

"Well, we have been eking along for more than a year now, but our earnings keep falling. We finally decided that we can't keep going any more so we decided to get out now and try to sell our stock and then the building so that we can salvage something from it," George said.

"But it always seemed to be doing well. You were always so cheery and willing to talk," I said.

"Yes, well, we had more time to talk than we wanted. We really weren't selling that much and we had to slowly let go our help. Then finally, we still couldn't make a profit," Nancy said with tears coming into her eyes.

Jenny came over and gave her a hug. "But how did this happen?"

"Once that MallMart moved in at the edge of town, back towards Indy, we just couldn't compete on the prices. Many of our customers went there for the lower prices. When we saw them or they came in for specialty items, they would apologize, but everyone around here is trying to make ends meet. If you can buy chicken for $1.35/pound over there rather than $1.75/pound here, people go for the deal," George said.

"We had a good long run. We were in that store for 35 years and I got it from my parents who started it in the 1950s. I hate to let it go from the family and be the one to see it decline, but we just can't compete with the big box store," Nancy said.

"We decided that we would retire and sell our assets if we can. We have enough to get by in retirement. Luckily, Nancy has been making us put a little away each month. We'll have to cut back, but it looks like the world is telling us it's time to retire," George said, also walking over to hug Nancy.

"We're just coming to clean things up. We'll have a few more deep sale days in the coming week to try to move the rest of our stock before we finally close the doors," Nancy said.

"Our kids, Joanne and Steven, seem to be taking it the hardest. They grew up in this store and thought that they would be running it once they got out of college. But it's just not making enough profit to support them anymore. We talked to them to let them know we'd be closing and they're heart broken," George said.

"We'll certainly miss you. We've been coming out here from downtown Indy for a few years. We found your place because it was in between our apartment downtown and my parents' house," Jenny said. "And we just fell in love with your shop."

"Stay in touch. Hopefully, we can come visit you once this is all settled," I said.

We climbed back into our car as Nancy and George went in the side door.

As we drove down the road, I remarked to Jenny. "I can't believe they're closing up. That was always such a great shop. Not too fancy, but always nicely kept with a good stock of groceries and unique items you don't find many other places."

"It's sad. I don't know another grocery like that in this area. I was envisioning our children growing up in this little town that is safe and away from the bustle of the city. Now I wonder if this little town will survive. It all seems to be due to MallMart moving in just five miles away and offering such cheap prices [2,3]," she said.

As we drove down the road, we could see other closed-up shops. The hardware store that we went to on occasion had a big Closed – Out of Business sign on the window as well.

Once this vibrant little town had its own small central business district. People in the town could walk to the grocery, the hardware store, the post office, and a coffee shop. There wasn't much, but it was most of what we needed around here. Now it looked like that was going away. As we drove through town, we passed the central square with the courthouse and the post office. Jenny had her hand on her stomach while we drove through the town looking at it with a new awareness of the health of the little town that we now called home. On the streets surrounding the square you could see the old grocery that still had an old-town look to it. The barbershop on the corner and the hardware store looked a bit faded and in

disrepair. It was obvious now that the town had been declining for a while.

Chapter 2

The Homestead

The Farm near Clay City, Indiana
April

Jenny and I drove over to her parents' place for Sunday lunch. We met with them a couple of times a month and it was nice to see them since they were only about an hour away.

We drove up a bit late. Lunch was already laid out and everyone was at the table. Jessica and Michael were Jenny's parents. They were spry and hardworking, but aging in their 70s. Chris, her brother, was doing most of the work on the farm these days. He was in his mid-30s, tall, blond, and strong. He had spent most of his life on the farm and was comfortable there, but he had seen the effects of climate change on the farm and was concerned about it.

"We saw on the nightly news that another corporate entity was brought down from the information that you published in Exposé on Climate Change, and how the board members of Extreme Oil Company were now under arrest for deceiving their stockholders and the general public about the problems with burning fossil fuels," Jessica said.

"Once the pieces were finally in place, it was too obvious to be ignored any longer. There are still a lot of lawsuits flying around and people are asking who knew what and when," I said. "There is a lot of finger pointing and heads are rolling on this one."

"Yes, but it's interesting that the CEO of Extreme came around and was working with the authorities," Jenny said. "That seems to be helping the authorities and the whole system switch over to a more climatically logical approach to energy acquisition."

"I'm glad that you pulled all of that information together. Even we were not convinced that climate change was happening. We had believed the fossil fuel industry propaganda. That really burns. I don't like being misled," Michael said.

"At least the truth is out now, and there's no going back. People are very much aware of what we're doing to the atmosphere. I even see the effect here on the farm and you can bet the other farmers are seeing it, too. Politicians can't lie about climate change any more. They have to face it now and make the changes that need to be made to get it under control," Chris said.

"Still our society is set on burning fossil fuels. It's the way of life that we've become accustomed to and at the rate we're going, we'll be lucky if we leave behind a world with just four degrees Celsius of global warming. That won't be pleasant: we'll have consistently warmer temperatures, shifting growing zones, and more intense storms," I said.

"Well, we're already adjusting. I've started to plant different crops that have done well in Kentucky and Texas. I even have a trial plot on the back acres with cotton to see how that does here [1,2]," Chris said.

"Yeah, I saw on the way in that you have a lot of experimental plots out in the fields. It's not just corn and soybeans anymore," Jenny said.

"Mom has been growing a whole variety of things in her garden, so I thought we could expand that to our production fields. Then I sectioned off the plots and I'm growing a variety of plants. Some are working and some are not, but it feels right to have this greater diversity of plants growing. We had a blight hit the corn last summer, but the tomatoes and cotton did really well. So it offset out losses," Chris said.

"That's a great benefit, but aren't their problems with changing from the main commodity crops?" I asked.

"I don't get as much subsidy from the government as I used to, but I'm happy not to be depending on the government so much. It's hard to learn how to deal with large scale production of these other crops, like tomatoes, but I like the challenge. None of it's easy, but it I feels like it's keeping me ahead of the changes in the market that could bring down the farm," Chris said.

"That's enough farm talk for now," Jessica said. "What're you working on these days, Greg?"

"I'm thinking of a story on sustainability," I said.

"What's that? Recycling?" Chris said. "That doesn't sound like much of a story."

"Well, that's the thing. Most people hear sustainability and think recycling or have no idea what it is. It's actually a complex mix of things from alternative energy, to food systems, to social systems, like healthcare, the minimum wage, and job security. It's having a healthy environment that we support and that also supports us with clean air and clean water. The problem is that we haven't viewed the system as a whole. We've taken parts of it, like coal, out of the ground to burn for energy, but we don't think about the system as a cycle and where that carbon will end up," I said.

"It's like farming," Jenny said. "When you grow the same crop on the same field year after year, that field gets depleted. So you can spread artificial fertilizer or you can plant something else like soybeans that help replenish the soil. And every few years, it's good to leave a field fallow so that it can heal itself. It's working with the natural processes to make a productive system."

"That makes sense. I remember studying ecology the one year I went to Indiana State University. They were talking about how the natural system is balanced and there's no waste in the system. The same can be said of agriculture when it's done well," Chris said.

"Why did you start thinking about this story?" Michael asked.

"As you know we moved to Robertsville a few months ago. We've been going to the same small grocery store every week. We just tried to go there and they're closed now. A MallMart was built just five miles away and now everyone goes there for the cheaper prices. It made me notice that most of the small mom-and-pop stores in our town are closing up. This big box store is driving all of the competitors out of business. It makes sense for people to shop there because it's cheaper, because they can't *afford* to be loyal to their old stores. They aren't getting paid enough in their jobs to decide to support a local store over choosing the cheaper prices. They're just living too close to poverty to have that luxury," I said.

"Well how can MallMart have such low prices and why are the local stores charging so much if you can get things cheaper?" Chris asked.

"MallMart forces its providers to sell to them at rock-bottom prices. Also, they buy in such large bulk across the entire country that they can buy at a much smaller profit margin. These are called economies of scale that allow the producers to ship larger amounts at lower cost to the big box store distribution centers and then MallMart distributes it to its own stores. All of this makes the costs very low and the store can succeed on very small margins. But the local stores don't have the luxury of buying in such bulk or arguing for the lowest possible prices. So they have to pay the true price for the objects and pass that along to their shoppers. In the end, they just can't compete with these big box stores," I said.

"These other stores, like MallMart, have a right to make a profit and do well in the business also. How can you argue against that?" Chris said.

"Yes, they do have that right, but their practices hurt the environment," Jenny said. "They push their providers so hard that the providers take too much from the land. So the end result is hurting the land. The producers also end up polluting the atmosphere and the earth with their production practices to keep costs low. These are called externalities. We all pay the price in environmental degradation, but it's not reflected in the price of the goods."

"We don't deny the stores the right to make a profit," I said. "They have played the business game well and are making a good market share for themselves, but there is a bigger set of problems out there. We are facing the loss of resources with peak oil. We have produced over half the available oil in the world and now we are on the down-hill side of that production. That's going to make fuel more and more expensive. We have already talked about climate change so you know how that threatens our lifestyle. All of that is being combined with the rich getting richer and the poor getting poorer. Our systems are out of balance. Regulations need to be put in place that provide a fair wage to workers. CEOs in the 1950s made 20 times as much as their workers and today they make 230 times as much [3]. The 1% of the richest people should not own 40% of the US resources [4,5]. All of these things are coming to a head, and we need to understand them. We need the system to come back into balance. These big box stores use people and use the environment to a point where it'll break the system. There need to be laws in place that require a well-managed environmental system that supports our livelihoods. Because without that, we won't have a healthy environment to pass on to the next generation."

"That's a lot," Michael said.

"I want to tell *that* story now. I think the people need to know about these environmental, social, and even profit systems that we depend upon. Then something needs to happen to correct the balance or we'll be harming all of human civilization," I said.

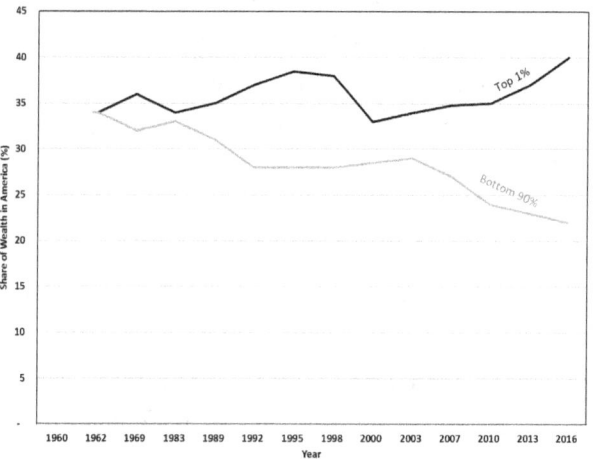

The wealthiest 1% of people in America now own 40% of the wealth while the bottom 90% of Americans own only 22% of the wealth. That also means that the other people in the top 10%, excluding the top 1% of wealthiest Americans own then remaining 38% of the wealth. [5]

We finished up lunch and Jenny and I helped to clean up the dishes as we continued our conversation.

"We've noticed a bunch of the old businesses around here closing up," Michael said. "We weren't sure what was causing it, but we also have a new MallMart, just down the road."

"We've gone there a few times, but it's such a big place and always so busy. I liked our usual places better. We still have the Rural King, but it seems to be getting less and less business lately," Jessica said.

"Yeah, I think this is a pattern all over the US. The small businesses are obviously losing out, but I think we are the ones that are really losing. We are losing our culture and the diversity of lifeways that used to be supported in a small town. Now everyone is strapped for money and just hoping the next big factory or the next big store comes into town to save things," I said. "But they don't save things. They just add to the demise. Most of the money that they make leaves the community and goes to their corporate headquarters."

"It seems the only towns that are doing well are ones that have pride of place and keep to their traditions. Those are now tourist destinations. People want to get back to the old ways, like the quaint old stores where you aren't sure what you're going to find. Also, places that you can walk into and know the owners and they know you. Places where you can talk and connect with people," Jessica said.

"Some places are bringing that back, but it is a hard battle," I said.

"The small farms are getting pinched as well. I know three farmers in the county that have had to sell their farms to big companies because they can't make ends meet. They couldn't afford the seed, fertilizer, pesticides, and herbicides this year to make the crop a reality. And that's not even thinking about renting the combine. I'm concerned that most the small farmers are going to disappear in the years to come," Chris said.

We had finished the dishes and were starting to pack up our things. Jenny and I made eye contact and I raised my eyebrows. She nodded a bit.

"So guys, on another topic. We've been wanting to tell you something but I haven't been able to find the right time, so I'm just going to blurt it out. We're going to have a baby!" Jenny said.

Jessica, put her hand over her mouth and started to tear up. Then she rushed over and hugged Jenny. Michael shook my hand since I was standing around awkwardly and Jenny was occupied with her Mom. Chris, came over and shook my hand as well and gave Jenny a hug.

"Congratulations! Do you know if it's a boy or a girl?" he asked.

"We don't know yet. We go to the doctor in a couple of weeks for the ultrasound that'll tell us," I said.

"We're so happy!" Jenny's Mom came over to me for a hug while Jenny's Dad went to hug her. "We've been wondering when and if we're going to get a grandchild.

You've been married five years. Of course, we didn't want to pressure you, but we aren't getting any younger. We always thought Chris was going to settle down and find a wife and get married, but it hasn't happened yet. I'm so excited to be a grandmother! "Have you thought about names?" Jessica asked as Chris looked a bit sheepish.

"We've been talking about it, but we haven't settled on anything yet. It just seems like such an important decision. I mean, whatever we name the child, they'll have to live with that name for the rest of their lives. It's a big decision," Jenny said.

"This is probably not the best time for this either, but since Mom brought it up, I guess I should announce it as well. Jane and I are getting married," Chris said.

"What! That's fantastic. You've been dating for a couple of years. I didn't think you were ready for that," Jessica said as we did hugs all around again.

"I just can't think of any girl that I get along better with. We've been friends for so long and dating a good long time. It's time that we made ourselves a family," Chris said.

"When's the wedding?" Michael asked.

"We're planning it for October. The weather won't be too hot or too cold and most of the work of the growing season will be over by then. So we thought it would fit in well. Jane also says it'll be nice with the fall colors in the background," Chris said.

"Great! Can I help with any arrangements?" Jessica asked.

"You'll have to talk to Jane about that. She'll be over tomorrow so you can spend time talking about wedding plans. We think we'll have it here. Maybe rent a tent and have it out on the back lawn," Chris said.

We packed up the car and headed down the drive, while Jenny waived to her parents and brother out the window.

Once again we drove through the familiar countryside. It was nice to be out on a warm Sunday in April. The greenish-yellow of emerging leaves gave a nice glow to the landscape. If you looked closely under the trees, you could see splashes of color from the daffodils, crocuses, and tulips in peoples' yards. Most of these were introduced flowers, but you could also see the carpets of white spring beauty and the smaller dots of yellow from trout lilies. These were the native spring ephemerals that came up early in the season as the trees were just leafing out. The air smelled fresh and all of the plants were getting a new start on the growing season. It was a lovely time of year.

Chapter 3

Selling the Pitch

Robertsville, Indiana
May 5

"Hey Charlie. How're you doing?" I said to my editor at the New York Times through the computer screen connected to Skype. I would often check in with him to catch up on the news and sometimes pitch a story idea, which I was doing today. I worked from home, sometimes traveled for research on a story, and Skyped in to talk to the office.

"I'm well. How're you?"

"Doing well. We're settled in to our new house. It's perfect for us. It's an old place that was built in 1860, but it's been well kept up. It has all of the original wood doorframes, window frames, and floors. They were never painted so they have a great natural wood color that has darkened over the years. It's a two-story house with four bedrooms and a basement. We have about 1.5 acres that are mostly wooded. It's beautiful and a nice place to relax. We don't have any neighbors that we can see from the house, so it feels very private," I said.

"Sounds like the perfect rural setting and a good place to write from. I've always been a city guy, but sometimes it feels like I'm constantly running from one thing to another," Charlie said.

"But cities are a great way to fit in a lot of people in a small space. The city can provide all of the necessary resources in close proximity to where people live in high rises and you can walk to everything that you need [1]. It is very efficient," I said.

"Yeah, I know. I've lived in the city my whole life. I don't even own a car. If I did, the parking fee would cost as much as your mortgage," he said facetiously. "And it's very hard to get around the city in a car because the traffic is so bad. The public transportation is much better."

"Actually, that's one of the best practices in sustainability and I wanted to do a story on sustainability. It seems to take many forms. Some people are talking about alternative energy, while others talk about permaculture, and still others focus on social capital. It seems to be a big thing, but poorly defined. Definitely most people don't understand it very well," I said.

"It's definitely a buzzword that you hear all the time, but it's vague in its meaning; probably because it means different things to different people," he said.

"Also, it's an old concept, but a new field of study. The University of Arizona didn't talk about sustainability at all when I was there or have it as a major. Now they're one of the big players in sustainability," I said.

"It definitely seems to be quickly developing. I do think that it'd be good to get a handle on that, and after your last piece, the New York Times is happy to look at whatever you want to pursue," he said referring back to my last major piece, Exposé on Climate Change.

"I thought I would make it part of a series and call it Exposé on Sustainability," I said.

"Well it's not very catchy, but it's recognizable," he said. "What're you thinking you need for resources and contract for this one?"

"I was thinking I could use a raise from my last story. A travel budget, although lots of travel is one of the problems our society has in becoming sustainable, and $40,000 for the delivered article," I said.

"OK. Like I said. The newspaper likes what you're doing lately, so go for it. Stay in touch and stay out of trouble," he said.

"Don't I always?" I said hanging up.

Chapter 4

Confronting MallMart

Kalamazoo, Michigan

May 15

I wanted to jump right into the investigation and see what I was dealing with so I made an appointment at the headquarters of MallMart in Kalamazoo, Michigan to see the CEO, Roy Anchorage. The building was a massive steel and glass structure. It dominated a city block. I had directions to go through the main entrance, take the elevator to the fifth floor, and find room 520.

I was surprised how little security there was here compared to the Extreme Oil Corporation Headquarters I had visited in Houston. Anyone was allowed to walk right in and in this case, take the elevator up to the top administration level to meet with the CEO. It was a nice change from the overwhelming security that I had experienced in Houston.

With all of the windows, you could get expansive views of the natural world at every turn. You could see multiple lakes out across the countryside with slightly rolling hills. It was very bucolic.

The interior design was modern but subtle. The chairs were a neutral grey color and looked comfortable to sit in. They had upholstered seats and wooden arms. The main desk was a massive wooden expanse with a secretary sitting behind it who wore a light headset plugged into one ear.

"Welcome to MallMart. How can I help you?" she asked.

I chuckled a bit to myself at the familiar greeting but in a different setting. "I have a meeting scheduled with CEO Anchorage," I said.

"And what is your name?"

"Greg Cunningham reporting for the New York Times."

"One moment please." She clicked a button on the console and said "Greg Cunningham to see Mr. Anchorage... Hm-hmm."

"Take a seat please. He will just be a minute," she said.

I looked through the brochures on the coffee table as I sat to wait. There were a number of brochures of stores opening up around the country. Each one looked the same and just the landscape changed a bit if the pictures included anything more than the asphalt of the parking lot. I did note with interest that a number of them had solar panels covering massive parking lots and most of them had skylights in the roofs.

"You can go on back. He will see you now," the secretary said.

"Thanks."

I walked back through the oaken doors and found his corner office with an expansive view of the city. The CEO was sitting behind a big glass-surfaced desk with his back to the windows. He stood to greet me.

"Hello Greg. It's nice to meet you."

"It's nice to meet you as well. Thank you for consenting to the interview," I replied.

"My pleasure. We like to promote the corporation as much as possible so I'm always happy to talk to journalists." We sat down around a smaller wooden table with two chairs.

"I saw in the brochures out on the table that you are using sustainable solutions of skylights and solar panels on a portion of your buildings," I said.

"Yes, we use skylights on all of our buildings. The natural lighting is better for peoples' moods and reduces the cost of lighting since some of it comes from those windows. Our other lights in the building are set to dim as the outside light increases to reduce our energy costs ," he said.

"That's a nice use of technology that provides benefit to people and saves the company money; it also brings in the social benefit making it a great example of sustainability," I said.

"The stores that you saw with solar panels are all in the American Southwest. There is so much sun that the panels create a lot of energy and the states have developed incentives for the installation of solar power. Also, we put the solar panels up on racks over the parking areas so that the customers get the benefit of shaded parking. This is extremely important in Phoenix when it's 120 degrees out in the shade in the summer time. It gives our stores an advantage because we are the only ones providing covered parking right now. And the panels produce a large amount of electricity for the stores. It's a great win for the company and the customer," he said.

"I like the use of the panels to provide shade. I've spent time in Phoenix and shade is definitely needed," I said.

"So, what is your article about?" he asked.

"I'm writing an article about small towns and how they can survive in the modern day," I said.

His brow furrowed a bit, but he jumped into the subject. "Well, Mallmart is a great support for small towns. We bring jobs and things that people need at affordable prices."

"It seems that you have a store in just about every small town and even in the outskirts of the major cities across the US," I said.

"We've had great success in expanding. Our business model of providing the best value has been effective and people love to shop at our stores," he said.

"What's the effect on the other small grocery and hardware stores in these small towns?" I asked pointedly.

"We improve them through our competition or we improve on them by providing a better service to the people in the town."

"It seems that your business model is to drive the local stores out of business," I pushed.

"That's business. If the smaller stores can't compete then they have to close. The consumers make the decisions about where they want to shop. They vote with their money," he said starting to get a little agitated. It appeared that he did not like the direction that this interview was going.

"But you have the advantage of your huge supply chain where you can buy in bulk. You force your providers to run down their costs in order to get a contract with you and those factors allow you to undercut the price of any local store," I said.

"Like I said. This is business. My job is to make a profit for this company, so I need to push a business model that enables us to grow even if that is at the expense of the small stores. It's just business; it's not anyone's fault," he said.

"But the playing field isn't even. You have too many advantages coming into the game," I said.

"Still the city councils want us to be there. They provide tax breaks and code variances when we want them. They're happy to have the jobs that we provide," he said.

"But your jobs are low paying, usually without benefits, and very often not full time. When you get the tax breaks you aren't even paying into the community to improve the roads and the sidewalks that you're using," I said.

"The city councils should think about those things before they offer the tax breaks. They're all too happy to have us come to town," he said.

"That might be changing. I think that the city planners are starting to notice how towns die and people end up leaving after you come into a place. It's like you suck all of the money and happiness out of a place when you move in," I said.

"Happiness, what is this hippy kumbaya nonsense?"

"The small stores provide community and connection. Our neighbors run them and you get to know them and the people that shop in the store. This is the connection that people need. Your mega-stores take that away from the community. People don't have enough money to decide not to shop at your store, but they lose a part of themselves as their community crumbles and all they are left with is one big MallMart that is the only place to shop in town," I said.

"What type of story are you writing? This nonsense that you are spouting is counter to the very capitalistic nature of our country. We have a right to make money. We need to compete with the other stores and it's our job to win that competition. It's no concern of ours if the smaller stores go out of business, that's just more customer base for us. Everything that you're saying just shows that we're winning!" he said with his voice raising throughout his tirade.

At that point the secretary walked in with a tall swarthy man.

"Is everything OK?" he asked in a deep voice that was thick with what seemed to be an eastern European accent.

"Mr. Cunningham was just leaving, Aabid. Can you escort him out of the building?" Mr. Anchorage said.

I got up and headed for the doors. As I got close to Aabid, I could see the bulge of a gun under his suit jacket. I guess MallMart is not lax on security, but is a little more subtle in how they present it.

We rode down the elevator with an awkward silence between us and he escorted me to the front doors.

I looked back as I headed to my car and saw that he was watching from the front entranceway with his arms crossed over his chest.

Chapter 5

The Sustainability Tour

Terre Haute, Indiana
May 30

I was still working on understanding everything that goes into sustainability, so I thought I would visit an old friend close to home. I had interviewed Dr. Seamus Flanagan at Indiana State University about dendrochronology and the Hockey Stick Curve. It turned out that he was also deeply involved with sustainability at the University. I called him and set up an appointment with him. He said that we should meet by the fountain on campus.

"Hello Dr. Flanagan," I said.

"Greg! Call me Seamus, remember," he said.

"Right, it's hard to do after all of these years calling professors by their title," I said.

"Well, I don't stand on ceremony and am happy for people that I know to call me by my first name," he said. "I wanted to meet out here, so that I could give you a sustainability tour of our campus. You had asked on the phone about clarification about what sustainability is, and the best way to answer that is for me to show you what we are doing with a sustainability tour around campus."

"That sounds like a great idea," I said. "I always like a good walk about."

"As you know, we are situated in the Midwest. Actually, Terre Haute is known as the Crossroads of America because of the Old National Road, Highway 40 that goes east and west along with Highway 41 going north and south through town. We also have the Wabash River on the edge of town and many major railway systems coming. The university campus is right next to downtown, so students can head downtown for something to eat, or a coffee shop. Everything is close by so that you could walk or bike to most of what you need around here. This walkability factor is important in sustainability," he said.

"I've always appreciated the resources a downtown area provides being next to a campus. It seems like a good arrangement. What has ISU done to improve its sustainability?" I said.

"If you look to the southwest over there, you can see the blue vertical-shaft wind turbine by Sycamore Towers. Those are the undergraduate residence halls. They are the tallest buildings on campus. In Fall of 2010, we had our 400 students in the Introduction to Environmental Science class hypothesize where the best location for a wind turbine would be on campus. Then we had them discuss methods for how to test their hypotheses. We ended up with a 50-point grid across campus that was rotated off the north and south direction, so that the points did not just line up along the roads. Then the students came out for five minutes of every lab period each day and measured wind speed at some of the designated points. When all of the data was collected, we found that our urban structures helped to funnel the wind and increase the wind speed right at that location. So we installed the wind turbine there," he said.

"That's a great teaching process to work with the students and I bet they felt ownership of the wind turbine once it was installed. Why did you go with a vertical shaft turbine?" I asked.

"We wanted a vertical shaft wind turbine to catch the winds from any direction and to fit better into our tight urban landscape. But we were a bit ahead of our time, because we could not source one from any US manufacturers. In the end, we had to get it made in China and shipped over," he said.

"That's a rather sad state of affairs if we have to buy our alternative energy technology from China," I said.

"I agree and it shows that our government hasn't been supportive enough of research and development into alternative energies. The next Rockefellers are going to make their money on alternative energy and sustainable solutions. I feel that the US has not been leading that charge," he said.

"Over there we have our Enterprise Car Share™ vehicles," he said. "We have two cars for a car share program. Any student with a valid driver's license can use the cars.

You pay a small annual fee and then pay $7.50 an hour after that for use of the cars. It's a good deal for everyone because Enterprise gets more short-term rentals, students don't have to maintain a car and can drive a new car right from campus, and it reduces the vehicles on campus and the vehicle miles compared to everyone having their own car. The best part about it was that ISU did not have to pay any capital expense up front. Enterprise did all of the work to set up the program for us."

"What are these fancy buildings with the glass fronts and all of that natural light coming in?" I asked.

"These are our LEED certified buildings. LEED stands for Leadership in Energy and Environmental Design that was developed by the US Green Building Council. Environmental and energy efficiency concerns are taken into consideration in the design, construction, and certification stages of the construction or renovation project for those buildings. They bring in many important factors for human wellbeing including natural light in 75% of the space," he said.

"That's great, but doesn't it cost more for all of that extra labor for the construction and certification?" I asked.

"It's more costly to build to LEED standards and it does cost to certify the buildings, but if it's done right, you can get more savings from the energy that is saved every year than what you pay into the construction," he said.

"How did ISU get into LEED construction on campus?" I asked.

"In 2007, ISU signed on to the American Colleges and Universities Presidents' Climate Commitment which is run by the not-for-profit group called Second Nature. One of the goals that ISU chose was to make all new construction LEED silver certified as part of that agreement. ISU has done even better than that, because they have made all new construction and all renovation LEED silver. At least that's what we shoot for. Sometimes when it's certified, it only comes in as certified, rather than with a silver rating," he said.

"That sounds like an interesting agreement," I said as we walked through campus towards the grassy quad.

"It was one of the first major efforts by universities to become more sustainable in their actions. They had around 700 universities and colleges signed on to the agreement at its peak in 2010 and now there are just less than 600 active members still going. We signed on early and are still an active campus," he said.

"What other sustainability certifications and organizations are there?" I asked.

"The Association for the Advancement of Sustainability in Higher Education, called AASHE for short, is the leading professional organization for sustainability professionals. They have the Sustainability Tracking and Auditing Report Systems, which is called STARS. That is the gold standard for sustainability record keeping for an institution. Our own report is close to 300 pages and includes everything from human resource practices, heating and cooling system efficiency, and obviously our carbon footprint," he said.

"Your campus is very beautiful with all of the trees, shrubs, and grasses. That has to count for something in sustainability?" I asked.

"We're a Tree Campus USA, which means that we put money and effort into our tree canopy. We have over 3,000 trees on campus and put a lot of effort into maintaining campus with a nice canopy. Most of the plantings are natives and the grounds manager uses very little herbicides, pesticides, or fertilizer. She is even particular about the salt that she puts down to help melt the ice in wintertime. Some salts are bad for the soil and plants while others can actually provide nutrients for plants," he said.

"I never knew that there were even choices in de-icing salts," I said.

"There are, and of course the ones that are better for the environment are more expensive. But our grounds manager finds that they are worth the cost, because they don't harm her plants," he said.

I noticed a swarthy guy sitting on a bench on the quad facing the fountain. He seemed to look away when I glanced in his direction. We walked on for a bit further and I hazarded a glance over my shoulder. He was gone. He could have just been a student, but I could feel the hair on the back of my neck standing up. I would have to keep an eye out for him or others like him.

"What is that area over there with the tall grass plantings?" I asked.

"Oh that's our green roof. Normally, you have a green roof a couple stories off the ground because it's on top of a building, but our green roof is on top of our server space, which is in an underground building. So our green roof looks like a normal patio," he said.

"What are the benefits of green roofs?" I asked.

"They provide insulation for the building below, so it does not need as much heating and cooling. Also, the soil collects and stores water, which the plants use in transpiration. That also helps to moderate the temperature in the building and handles excessive water. It is a very efficient roofing material and is aesthetically pleasing," he said.

"It looks great. I wonder why more buildings don't do this?" I said.

"Well, it is more expensive than a normal roof and you need the structural support to handle the soil and the plants. But many places are starting to adopt this. Los Angeles has installed many green roofs which help to moderate their urban heat island effect and they provide green space in the middle of the city, so it's a win-win situation," he said.

"What is the urban heat island effect?"

"City centers are warmer than the surrounding countryside because of the buildings trapping hear, the construction materials that store heat, covering up soil and removing vegetation that reduces the transfer of heat away from the city through latent heating, or the evaporation and condensation of moisture. All of this adds to the heating of cities, called the urban heat island."

"What's next on the tour?" I asked.

"If we cross Cherry Street just into the downtown, we can check out the Scott College of Business, which is in the old Federal Hall. This used to be the US Postal Service building along with other federal offices. The US government sold the building to ISU, which has renovated the building. This was our first LEED Silver building and it was a renovation. They had to keep a large portion of the original structure and work around the thick limestone walls," he said as we entered into the building.

"These old mail boxes and all of the original fixtures are amazing," I observed.

"They spent a lot of time maintaining the original look of the hallways even to the point of keeping many of the original mail boxes," he said.

"They did modern improvements as well. I really like this computer classroom with all of the stock displays and the electronic ticker of stock prices. That looks like a great classroom to learn in," I said.

"It's a nice combination of old and new for sure. Come see this up here," he said as he headed up the stairs.

We went down a second floor hall to a large room, which was open as they were setting up for an event.

"That large mural is a depiction of the signing of the Magna Carta in England. This is the original federal courthouse and that mural used to stand behind the judge's bench. They worked to restore the original chalk drawing," he said.

"It's beautiful. It must be two stories high," I observed as we walked around the formal room with its grey metal light fixtures and the two stories of windows opening up on the downtown area of Terre Haute.

We left the Scott College of Business building and walked up 7th street.

"University Hall is our first building built to LEED standards. It was before we had committed to being LEED Silver and was our first foray into sustainable construction, but it's one of my favorite spaces on campus," he said. "Our administration figured out quickly that sustainable construction like this improves our human experience in the building and also saves money in the long-run. So this was one of the quickest sustainability initiatives that we adopted. It is now best practice in construction to follow something like LEED construction techniques. It makes places that people are happy to work in and it saves money through the heating and cooling of the building after about seven years of use."

"That's a good return on investment if you make your money back in seven years. Then your savings on heating and cooling just come back to the University after that," I said.

We walked into the building with its high ceilings, wood floor, and modern painted walls. Down the hall and around the corner we entered into a four-story atrium space in the center of the building.

"Wow, this is an amazing space," I said a little too loud as my voice carried across the open space.

"An old dreary outdoor garden courtyard used to exist here in the center of the building with no access. The architects glassed in the roof and brought in 30-foot tall trees in planters. It is one of the preferred spaces on campus now. It has a nice coffee shop and is full of natural light. Many students come here to study. This building used to be the old lab school where kids came for preschool and the lower grades. The students learning to be teachers worked with the children for a practical application of their skills. The building was abandoned from that use years ago and was mainly used for storage. It had water damage and was quite a wreck. A lot of time and money went into this renovation, but now it's a gem of our campus. In my mind, the best qualities are the sustainability features that LEED certification suggests," he said.

We exited the northwest side of the building and walked farther up 7th street.

"See that clock tower?" Seamus asked.

"Yes. That's a nice building. What is it?" I asked.

"Its our main boiler facility that provides steam and heat to the whole campus through underground tunnels. Just to the south in that parking lot is where we used to have a coal-burning boiler. The old boilers were at the end of their life and the university president at the time decided to invest in a new natural gas boiler. The transition happened in 2002 and I was teaching a conservation and sustainability class at the time. I used to bring the students over for a tour of the coal facility and the natural gas facility. The difference was night and day. The coal facility was extremely dirty and hazardous working conditions. The windows inside the building were completely black with coal soot. The new natural gas facility is clean and bright; you can walk through and enjoy the space. That conversion from coal to natural gas halved our carbon emissions in 2002, so now we produce about 50% of the carbon emissions that we produced in 1990. That's a great improvement for the environment and is the best improvement that we have had on our carbon footprint analysis," he said.

We walked out of that building and headed back south and east towards 9th street.

"Our recycling facility over here came online in 1990 because we were paying too much in tipping fees at the landfill, so the university president decided to start a recycling program and see what we could divert from the landfill. The recycling center was so successful that the administration decided to open up the recycling drop off to the public. Just about anything that is recyclable can be dropped off here and its the only facility left in town that still takes glass," he said.

"Why's that?" I asked.

"It turns out that the input materials for glass, such as silica sand, are so cheap, that there isn't much of a recycling commodities market for glass. Therefore, when the center ship out a semi-truck full of glass, the center gets paid $20, which doesn't even cover the cost of gas. Our recycling center is still dedicated to reducing the amount of material that we send to the landfill, so they continue to recycle it at a loss. I was surprised to find out that the plastics commodity market crashed as well when oil was down to $20-$30 a barrel of oil. Plastics are made from petroleum. It wasn't worth the cost of melting down old plastics for their petroleum uses compared to buying fresh oil. But as the price per barrel of oil goes back up, the plastics recycling market will pick up again as well," he said.

"I never knew that the commodities markets for recycling were so complex and reliant on other external forces. I always thought that you just made money by selling recyclable materials back to the industry," I said.

"It's a very complex set of markets with each recycled item carrying its own price that is dependent on different market factors. Also, the grades of plastics matter and contamination from other sources is an issue as well, such as food on containers. Our recycling program produces high grade recycled materials, because they have the community members dropping off their products sort everything into different bins. That way, the recycling center can make more money off the higher value materials such as aluminum cans, white paper, or cardboard. They still collect the recyclables that don't make as much money, like plastic, right now and glass, but we hope those markets go back up in the future," he said.

"Do you think that all of the recycling commodities markets will be going up in the future?"

"It actually doesn't look very good for many recycling markets. China used to be the main destination for recycled materials. They would take any material and use it as a source for their industry, but it carried a high pollution load and wasn't good for the people that processed the materials. So, China recently decided to stop taking the world's recycling materials [1,2,3]. This has caused a large disruption in the recycling commodities market and we're not sure who will be taking these recycling products in the future," he said.

We walked out the back of the drop-off zone and down to Chestnut Street where we headed east.

"One aspect about sustainability that we haven't talked about is a little more hidden. It's our human resource practices and how we treat the people that work on campus. A major part of sustainability is social justice and that includes fair working practices as well as making sure that minority communities don't bear the greatest burden of poor environmental conditions, like landfills, in their communities. ISU has a policy of paying more than the minimum wage so that people can make a living wage. They have a good benefits package for their full time employees and work to keep them employed at the university. All of these factors play into sustainability as well," he said.

"Those are areas I don't usually think about," I said.

"The more obvious cases can be seen through the research of our students. We have had students map out the quality of the sidewalks around town and they found here, and consistently find in other cities, that places in the town with larger minority populations have worse sidewalks. This is because the often white and male city council members are making decision about where to spend money on improvements in the town and that money is often spent in places that they see and experience, so the minority population areas are often overlooked when it comes to street and sidewalk repair. Another case is when you map landfills across the country and you map the concentration of minority populations, the two are highly correlated. Again, this is from a series of decisions through time that placed those landfills away from the people making decisions which landed them directly in the neighborhoods of highly diverse communities," he said.

We crossed train tracks on a slight rise and stopped for the little view that the rise gave us.

"These train tracks are a symbol of the past productivity of this town," Seamus said. "As I mentioned before, Terre Haute is known as the Crossroads of America. It's on the Old National Road, which is Hwy 40 going east-west across the US. It also has Hwy 41 cutting north-south through town which connects Chicago to Miami. We also have many railroads that cross through the city. These came in the 1860s and were the main means of industrial transport since that time. They are still quite active with more than 30 trains a day just on this one track. We used to have a passenger train that would go up to Chicago every day, but that route stopped in the

1960s."

"That is a lot of modes of transportation," I said.

"That's not all. Before the highway and railroads came in, we had the Wabash and Erie Canal that came through here and was very active from about 1840 through 1860. That system went bankrupt when the railroads took over. And before that, we have the Wabash River coming through town. This was a major transportation corridor since the earlier European settlement of the area in the 1700s and it was extensively used by Native Americans before that," he said.

"That's a lot of crossroads. I'm surprised that Terre Haute is not larger than its 60,000 people today."

"Well, people hypothesize that it was actually prohibition that set the town back. This region was the major brewing and distilling center for Indiana in the early 1900s. It was a main hub for grain coming in from the fields and also a large livestock area that focused on pigs and pork packing to be shipped out around the Midwest. But with prohibition, much of its industry shut down. Before that time, Terre Haute was larger than Indianapolis. But after that, Terre Haute started to decline and Indianapolis grew and became the dominant city in the state," he said.

"That's an amazing history for this small town," I said.

"Most small towns have an interesting and storied past when you look into them. We just don't usually take the time to understand the history of an area," he said.

"What is that area over there with the high fence?"

"Oh, speaking of our history meeting the present. That's ISU's oil well!"

"What? You're drilling for oil on campus?"

"Yep. Our last president was a petroleum chemist and it turns out that this was a big oil and coal producing area in the past. The small town of West Terre Haute was established as a coal-mining town in the mid-1800s. There used to be eight oil well heads in the downtown area in the 1800s. Most of those are now gone or capped, but we still have oil bearing geologic layers deep underneath us," Seamus said.

"How deep?"

"Down about 1500 feet you hit the Niagara Limestone which is oil bearing. We have had a series of small oil derricks pumping around town for some time. Our university president looked into the potential of these rock formations under the University. It turns out that the University holds the mineral rights, so he decided to explore oil drilling on campus with new technology of horizontal drilling."

"Is that anything like fracking?" I asked.

"No. Fracking or hydraulic fracturing is when a company breaks the rock units and pumps in a slurry solution to force out natural gas. This is traditional liquid petroleum that is being pumped out.

"When did the oil well go in and how did the University community feel about that? Most schools are divesting of the petroleum industry, not getting into drilling for oil on campus."

"It resulted in one of the few major protests by students that I am aware of on campus. We only found out about it because of a Tribune Star story on a Wednesday that stated the Board of Trustees was going to vote to approve exploration and drilling for oil underneath campus on Friday. This was back in 2013," Seamus said.

"I bet that didn't go over well," I said.

"It did not. The student Environmental Club organized, as did the Campus Republicans, of all people, to protest. I was able to meet with our campus president just before the board meeting and state our concerns. For one thing, we had just recently signed on to the President's Climate Commitment where we committed to be carbon neutral by 2050. This seemed to be the death knell for that agreement," he said.

"Was it?"

"No. I talked to Second Nature and they stated that the carbon cost of the petroleum is counted where it's burned. We only had to account for any methane that escaped at the wellhead. This well was drilled in January of 2014 when I was on sabbatical. Of course, they located it on the periphery of campus so that it was not so obvious, but it's also located 50 feet from our sustainability center," he said a bit incredulously.

We walked a little further along and came to an expansive chaotic garden.

"This is the ISU Community Garden. In our garden, anyone in the community or the University can get a plot for free and grow their own food. Many community gardens charge a nominal fee for the use of the plot and to maintain tools, but ISU decided not to charge for plot space. Instead, they asked their gardeners to give ten percent of their produce to a food bank. ISU provides water and repairs or replaces the tools as needed. So the gardeners just need to bring their plants, their ideas, and their energy. Each gardener can get a 10-foot by 10-foot plot. If they prove themselves by maintaining their plot and the local community areas, then they can get more plots or larger plot spaces. I think one gardener has three 20 X 20 foot plots now. She is a master gardener and has been here since the garden started," he said.

"I can see the different plot sizes laid out. You can tell the different plots by the variety of things that are grown in them and the variety of ways of growing in the plots," I observed.

"Yes, we have about 175 gardeners and about that many plots. Each person comes with their own idea of what they want to grow and how they are going to work on their plots. We also have a very culturally diverse set of gardeners. We have people from about five different countries in Africa, from China, India, Vietnam, Sweden, and Germany. Each person brings their culture and tastes with them for what they grow, so it's quite an education just walking around the plots and observing what people are doing," he said.

"What is that house with the great deck?" I asked.

"That's our Institute for Community Sustainability. Let's go inside," Seamus said.

"This is a nice little office space," I observed as we went in.

"There were eight houses on this half block before the garden came along. All of the houses were abandoned and run down. So Indiana State University bought up the lots and tore down the houses to create a buffer of green-space around the campus. Later this area was developed into the community garden," he said.

"Wasn't that old housing stock and land use an issue for the quality of the soils?" I asked.

"Actually, it turned out to be a serious concern. We purchased a portable X-Ray Fluorescence instrument that allows us to measure 39 elements in the soil in about 90 seconds. We tried it out here the first time that we used it and found elevated heavy metals in the soils, especially lead. After that, we had an undergraduate that was interested in the project take it on as a senior thesis project. He systematically took nine samples from each plot and mapped out the lead contamination," he said.

"That must have been a lot of samples."

"The student had over 1,000 samples by the time that he was done. Luckily, the analysis on each sample was quick. We found elevated lead and arsenic levels in many of the northern plots."

"What did you do about the gardens in those areas?" I asked.

"We worked with the gardeners and educated them about safe garden practices. Only a few of the plots had really elevated levels and it was a spotty pattern. Much of our work since then has become about remediating a site with lead contamination [4]. We covered most of the high contamination areas with raised beds because the lead is not very mobile and is not picked up by the plants. So sealing it off and not digging or tilling in the soil was the best remediation that we could provide and it was also a cheap solution. Now the garden is a good example of how to remediate an urban area with documented lead contamination. We continue to have our students testing the lead in the soils every year so that we can track it and see how it changes through time," he said.

"That's an interesting story and I imagine a lot of these areas that are now doing urban gardening could be affected by heavy metal contamination," I said.

"We found that in Terre Haute, it was really an old building signature. We had a master's student continue the lead project. She took another thousand samples on a coarser scale that covered the entire City of Terre Haute. She found the areas with houses built in the 1800s had the highest soil lead contamination. The houses around the garden here were built in the 1920s. But for all of that time, lead was used in paint because it made the paint more pliable. So over that time, a lot of lead paint was put on the houses. As the paint aged after decades, the paint flaked off or was sand blasted off for new layers of paint and all of that lead went into the soil where it has remained until today. This is the reason that Vigo County has one of the highest incidents of elevated blood-lead levels for children in the United States," he said.

"I never knew that was such a problem. Now I'm concerned about our old house. Maybe I'll send in soil samples for your lab to test," I said.

"That would be fine. We actually provide free soil lead testing for the community. Also, most public schools in this area do regular blood lead testing in children. That is how we know we have an elevated number of children with lead poisoning compared to the rest of the country," he said.

"What can be done about it?"

"Well, the most important thing is to get it diagnosed. It's possible through chelation to remove the lead from blood. It's not an easy process, kind of like getting multiple blood transfusions over time. But if it's left untreated, it can lead to neurological development issues, learning deficiencies, and even excessive anger and violence in kids if it's very bad," Seamus said.

"It does seem necessary to treat it then. I can see where this's tied to sustainability. Past land use that was a normal process resulted in lead contamination in peoples' neighborhoods."

"Yes, a lot of our behavior that seems normal, results in contamination and deterioration of the environment. We don't think through the long-term unintended consequences of our actions. It's actually hard to do that. We focus on the one short-term gain that we are aimed at, but that leaves us blind to the other consequences. And there always seem to be other consequences," he said.

"So how did the Institute for Community Sustainability come to be in this house?"

"We kept this house because it was in the best shape out of all of the houses on the block. The interior was in disrepair so we could renovate it as we wanted with a small budget. We decided to make it a model home for sustainable residential solutions. The flooring is bamboo, which is a quickly replacing wood type product. The walls are painted with no Volatile Organic Compound, also known as no-VOC, paints. The cabinets in the kitchen are my favorite part. We decided to make a small teaching kitchen that could also be used by the people in the house. An Amish carpenter built all of the cabinets from hickory that was harvested in land just about 45 minutes to the north of here. He came and took measurements, handmade the cabinets, and then came down with his son to install them. He used every bit of area in this small space. The cabinets are full of little nooks to hold spices, a cutting board, and fold out shelving to use the full area of the corner cabinets," he said.

"This's a really nice kitchen. I wish I had something like this in my home," I said.

"It was a great use of the space and the hickory wood is a beautiful color. All of the appliances are highly rated Energy Star™ appliances that use little energy. Even the glass knobs on the cabinets are made from recycling windshield glass, which is hard to recycle. A Catholic Charities group in the Pacific Northwest make the knobs out of old windshields and upcycles them for a profit," he said, obviously proud of the beautiful kitchen.

"What're these weird ceiling fans," I asked looking up.

"These are called Sycamore Fans and use biomimicry to improve their design. They're a single-blade aluminum hollow-core fan. They use only 25% of the energy of a typical ceiling fan and work very efficiently. They are designed after the shape of a maple seed, often called whirly gigs around here, because of how they helicopter down to the ground from the trees when the wind blows. They mimicked the shape of nature and found a more efficient way to move air with the fans. They help to keep the house cool in the summer and push the warm air down from the ceiling in the winter so it reduces costs for heating and cooling," he said.

We walked out the back door onto the back porch.

"This's a 40-person teaching porch where we can hold classes outside. We received a $100,000 grant from the Lilly Endowment to build this along with seven passive solar greenhouses in the Wabash Valley. You can see the greenhouse over there on the northwest side of the garden. The greenhouses don't take any external heat, but the large south-facing glazed surface allows light in so the temperature is warm enough to grow throughout the fall, winter, and spring. It actually gets too hot to grow in the summer time. The glazing on the south side is a double layer of plastic that is inflated with a small fan. This provides great thermal protection for the plants while letting in the shortwave radiation from the sun," he said.

I admired the lean-to type structure that had a large growing capacity. "How do you keep the heat in the building?" I asked.

"We have 20 barrels that hold about 50 gallons of water each. The black barrels take in the sunlight during the day and heat the water. The water reradiates the heat at night keeping the greenhouse warm all night. It's a great design from the University of Missouri extension office. We've six other similar greenhouses in the Wabash Valley. Each was built with slight modifications and they are used in different capacities at each institution, so over time we can compare what worked and what didn't to learn how to make more efficient greenhouses," he said.

"This's a great place and it's nice to see all of the produce being grown right here in the middle of the city," I said.

"We can provide a lot of food in a pretty small area through the diverse plantings that we have from the different gardeners. We're also starting to record the biodiversity out here and find that it's quite high for the county as a whole," he said.

"I wouldn't have identified many of the things that you talked about on the sustainability tour as sustainable solutions but now I can see how they all work together to create a more sustainable future. I can see how it all fits together," I said.

"All of this and more is sustainability. It's really about making balanced functional systems that respect the people, the planet, and profit. If we fail at any one of those, it'll not be a sustainable solution."

"This'll help me to think about the breadth of ideas for the rest of my article. It was good to talk to you again," I said as we shook hands and I headed back across campus to my car.

Chapter 6

Peak Oil

Kinsale, Ireland
June

I needed to understand the pressures that were facing our society, so I travelled to Kinsale, Ireland where the Transition Town Movement began [1]. Kinsale is a lovely harbor town on the south side of Ireland [2]. The sea blows in cold and brusque with the salt spray coming off the ocean. I was meeting with Davie Wilson to talk about the Transition Town movement at a gastropub in the center of town.

I walked into the warm pub on a cold wet evening and was greeted by the sweet smell of burning turf coming from the fire in the corner. The walls were dark wood paneled and decorated with the memorabilia of days gone by. They had pictures of past patrons and bar tenders, a bodhran Irish drum, and a dartboard. The bar was a prominent feature with an exhaustive section of Irish whiskies; the smell of fresh baked bread and Sheppard's pie wafted from the kitchen. I recognized Davie sitting at a table near the fire. He was shortish with a black beard, black shirt, and a vest over the top.

"Hello Davie, I'm Greg Cunningham," I said by way of introduction.

"It's good to meet you. Have a seat. How were your travels?" Davie asked with an Irish accent.

"They were good. I haven't been to Ireland before, so it's nice to be able to come and visit. I believe that some of my family is from Ireland and I always wanted to travel here," I said.

"A lot of folks from around the world have Irish heritage. That comes from the potato famine in the 1840s when about 20% of the Irish population moved to other countries. A large number of them moved to the United States, which was booming at the time[2]," he said.

"I wanted to talk to you about the Transition Town Movement," I said. "It started here in Kinsale, correct?"

"Yes, I was teaching at Kinsale College and it was a natural outgrowth of my experiences in architecture in developing countries, and my realization of multiple societal pressures that're occurring at the same time," he said.

"What are those societal pressures?" I asked.

"Well, I know that you're quite familiar with climate change. I read your *Exposé on Climate Change* and saw the fallout from that in the news. Climate change is one of the major pressures on our society. It's a growing concern that will be with us for at least the next 100 years and that's if we act now to curb our carbon emissions. We're changing the atmosphere that we depend upon to survive," he said.

"I completely agree. I just hope that we can make the necessary changes to reduce our carbon emissions to avoid the worst of what the scientists are predicting," I said.

"Connected to that is the concept of Peak Oil. M. King Hubbert was a petroleum geologist for the Shell Oil Company. He examined the rate of discovery of new oil reserves from industry data and noticed that rate of discovery was dropping off in the United States. In 1956, he predicted that the United States would hit the peak in its production of conventional oil around the 1970s [4]."

"What is conventional oil?"

"It's oil produced from readily accessible oil wells on land compared to harder to reach oil reserves that take a lot more investment and energy to produce. His prediction was accurate and the US production of conventional oil decreased after about 1974," he said.

"Was this just a phenomenon in the United States?"

"No, Hubbert made predictions for many countries and even extended his predictions to global oil production. Of course, he admitted that the data sources were not as good for many other countries like Russia and Saudi Arabia that did not want to share their oil recovery and exploration data with other countries. History shows that over 60% of countries have passed their peak production of conventional petroleum," he said.

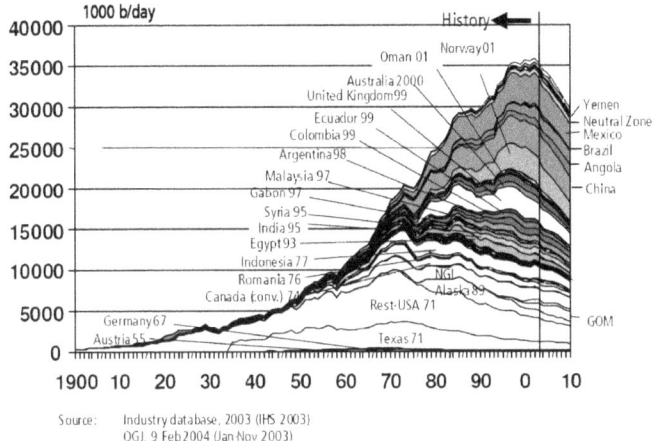

Source: Industry database, 2003 (IHS 2003)
          OGJ, 9 Feb 2004 (Jan-Nov 2003)

Hubbert's curves for each oil-producing country. According to these numbers, we passed global peak oil around 2004 and have less than half of the oil's original oil reserves left in the ground (graph created by the US Department of Energy). [4]

"What would you call unconventional oil?" I asked.

"That's an important question. In the 1950s, Hubbert was only aware of what we call conventional oil reserves today. So his predictions only accounted for those sources of petroleum. Unconventional oil, on the other hand, would be oil that is extracted from oil sands, drilled at the bottom of the ocean like in your Gulf of Mexico, or drilled from Polar Regions that are recently free from ice. These are areas of oil reserves that used to be too expensive to mine. Now with improvements in technology and the need for other oil sources, companies have found it economical to mine these harder to reach oil reserves that could be called unconventional," he said.

"Isn't that much more costly?" I asked.

"It is. It makes me think of your US TV show called the Beverly Hillbillies. In that show, the lead character shoots his gun, accidently hits the ground, then crude oil starts to pour out of the bullet hole, and they have found the first oil deposit. That's not too far from the beginning of oil production in Texas. Oil was so easy to drill, that a simple borehole would often result in oil flowing out of the ground under pressure. Now we have to start drilling 5000 ft under the ocean's surface to extract oil from the bottom of the Gulf of Mexico or dredge up a huge part of Alberta about the size of South Carolina to boil out oil from the oil sands. This means that it takes a lot more energy to produce a similar amount of oil," he said.

"But this unconventional oil has increased oil production?" I said.

"Yes. Because of these other oil sources, we see a departure from Hubbert's predicted curves for the first time. Some people point to this and say that it shows that Hubbert was wrong, but I believe that it actually supports Hubbert's argument. We are past peak oil that could be obtained by conventional means for almost every country. By innovation, we have found ways to extract the little oil that is left in hard to reach places. In the end, that just pushes back the date of complete failure of the oil industry. But as you saw with your climate change article, the oil companies will do anything to make a little more profit on this outdated technology," he said.

"It also seems that this fossil fuel production is in a feedback loop with climate change. As we produce more fossil fuels, we drive more climate change. Then we have to run our air conditioners longer, at least the people with the economic luxury to have air conditioners. This increases our carbon emissions and warms the climate further," I said.

"It's a vicious cycle. That is not even touching on coal. All scientists agree that coal is plentiful. The United States has about 300 years of coal reserves based on current usage rates. However, coal emits more carbon and produces more harmful pollution such as lead, arsenic, and mercury. Many countries, like China and the United States have large coal reserves, but that industry is still dying because natural gas is so much cheaper than coal. The coal industry has been declining for decades, but your President Trump campaigned on the idea of bringing back coal," he said.

"I know; there are many more jobs in solar or any other alternative energy than there are in coal, but our politicians seem to focus on the need for coal and the jobs that it provides[5, 6]," I said.

"Then natural gas also comes with its own problems. Its production has increased because of new technologies like hydraulic fracking, where the mining companies fracture the rock deep underground and inject a hot solution that forces up natural gas and oil. This has produced booms in production in natural gas like whole towns that have emerged in North Dakota to support the natural gas industry. But it comes at an environmental cost. Peoples' well water has been contaminated with methane [7] so that the taps can be lit on fire. Methane has been found to seep out of the ground and cause cattle to die. We are getting more cheap energy, but at what cost?" he said.

We got another pint of beer and continued our conversation.

"The Transition Movement is made to deal with these problems," Davie said. "If you follow peak oil and climate change to their logical conclusions, that means that we will need to live a local life. The loss of fossil fuels will mean that we won't be able to jet around the world, or even drive across town any more. Climate change also means that we are going to have to reduce our carbon emissions to do our best to mitigate the worst climate change consequences. All of this means living a more local life."

"I find it interesting to have this conversation when I just flew over from the states," I said feeling a bit guilty.

"Yes, that's true. We don't encourage people to travel all this distance to see what we've done. We hold webinars so that we can talk to people in their home communities and give them ideas on how to naturally reduce their carbon consumption. We work towards developing Energy Descent Action Plans with local groups in their communities to reduce their use of fossil fuels. Many of our members have committed to a no-fly agreement to try to reduce our carbon emissions," he said.

"That would be hard for me. I fly to many of the places that I study and having those in-person interviews and seeing the places where these things happen make it more authentic for my readers and me. But I understand where you're coming from. I do feel bad for all of the carbon that I've burned flying around to tell the story of climate change and now sustainability," I said.

"That's a tough decision. Travel, global experiences, and appreciation for diversity are often times at the core of our concern for the planet and at the core of how we interact with the Earth, but at some point we need to all stop burning so much carbon and act more locally," he said.

"What are examples of local action?" I asked

"In Totnes, England we have developed the Totnes Pound. This is locally printed money that can only be spent in the town of Totnes. The beauty of it is that you know the money will stay local for the benefit of the local economy and your friends and neighbors. From there, you really commit to act locally and improve your area. This, along with the no-fly agreement, really makes you focus in on the local community and put all of your effort into improving that community," he said.

"The town that I live in now can really use that. We have large corporate stores coming in to our area and it's driving the small stores out of business. They just can't compete with the prices at the big box stores."

"But I bet they do provide a better shopping experience and you feel connected with the community when you shop at the local stores," he said. "That's the wonder of this movement. It actually gets us to reconnect with our neighbors, to value community, and improves our daily lives more than anything else I've experienced."

"I can see that. I've read about how we're missing our connections with each other and with the community, which reduces our collective happiness," I said.

"It really is the connection with other humans rather than wealth or endless work that makes us truly happy. It's such a basic thing, but often times it gets overlooked as we strive to succeed in life in other ways," he said.

I stayed in Kinsale and explored the town and some of their efforts to improve the local community. After that, we travelled by train to Wexford and then took a ferry to Wales. Another train took us down to Totnes and we spent some time exploring this small English town and it's efforts to stay local.

Chapter 7

Our Healthcare System

Aviemore, Scotland

June

I had been running hard for a while. Ever since I got the contract for the story exposing the oil industries' role in climate change, I had been flying all over the world to talk to researchers to investigate. I was travelling less now as I researched sustainability and I realized that my carbon footprint was so drastic that I reduced my travel. This was a rare trip to the British Isles for me so that I could interview the people that started the Transition Town Movement and I felt it was necessary to meet with them face to face. Now I had some time on my own and wanted to appreciate the region before I went back home. So after my interviews with the folks in Ireland and England, I travelled to Aviemore for some relaxation in the Scottish Highlands. I took the train through London and up to Edinburgh. After changing trains, I traveled into the highlands. It took about 12 hours of train travel, but I could cover the entire length of England and Scotland on train. I found train travel to be easy and comfortable. I could work at a table, plug in my computer, and even got wifi for some of the stretches. I worked on writing what I had so far for my sustainability article as I watched the English countryside pass by. I made my way to the Mac-Donald Aviemore Highland Hotel and stayed in one of the lodges back in the woods. It was a peaceful place with trails into the nearby mountains and a large campus. I had been under much stress lately and thought it would be good to relax in this out-of-the-way place for a few days.

As I got settled in for the evening, I thought I would have a nice read by the window and just unwind after my travels. I found that I was agitated and couldn't settle down. I felt a pain in my chest and my arm was tingling. My breath was short and I felt a bit dizzy and nauseous. I sat for a while, hoping that it would pass. When it continued for a while longer, I thought it would be best to call for help.

I dialed 999 on the hotel room phone, which is the emergency response number in Scotland. A kind voice answered the phone and asked with what they could assist in a singsong Scottish accent. I told them that I was not feeling well and could be having a heart attack.

They told me that assistance would be with me in a few minutes. They took my location information and told me to take aspirin if I had some.

The ambulance arrived in about 5 minutes. They had me lay down on a gurney with collapsible legs that they had brought up to the lodge. They checked my vital signs.

"We would like to take you out to the ambulance where we can do some more tests. Is that OK?" one of the paramedics asked.

"That's fine."

They rolled me out the front door and carried the gurney down the couple steps to the path. When we got to the ambulance, they pushed the gurney into the back of the ambulance. They closed the doors and hooked me up to an EKG machine with adhesive pads stuck on my chest and even my ankles.

They monitored me for a little while.

"Your EKG is not bad but it is inconclusive. We would like to take you to the hospital for some bloodwork and to be monitored for a few hours," the paramedic said.

"OK."

One person went around front to drive, while the other stayed in the back with me to monitor my vital signs as we drove through the quiet streets late at night to the local hospital.

As we drove, the paramedic took down my name and contact information. I had bought a Scottish SIM card at the airport so I had a local phone number. Other than that, I could just give them the name of the hotel where they picked me up. She wrote all of that down on a small form. She also asked some questions about my medical history and wrote down my answers.

It took about fifteen minutes to get to the hospital in Inverness since we were so far out in the highlands. Once I was at the hospital, they pushed the gurney out of the ambulance and into a private room. A doctor came and checked on me. He was holding the form that the paramedic had filled out.

"Hello Mr. Cunningham, I hear that you are experiencing some chest pain," the doctor said.

"Yes. A tightness in the chest and my left arm hurts."

"We are going to do some bloodwork, but your EKG looks good."

A male nurse came in a few minutes later to draw some blood. "Hello, my name is Chuck. This will just hurt a bit," he said as he inserted a small needle in my arm and drew out two vials of blood.

We chatted while he worked.

"When do I fill out the paperwork for payment for all of these services?" I asked.

He chuckled a bit, "You're in Scotland now. We have universal healthcare so you don't have to fill anything out or worry about payment. I assume that the paramedic filled out a form on your medical history on the way in?" he asked and I nodded. "That is all the paperwork that you will see here."

"Really? If I were in the United States, I wouldn't even be treated until I could demonstrate that I had insurance or could pay for the procedures [1]. These tests and the time in the hospital and ambulance ride must be expensive," I said.

"Not really. Because it's paid for through National Health Services with the country, it keeps the costs down. Everyone is expected to get good health care and everyone pays for it with his or her taxes. It keeps the costs of the entire system under control and provides basic coverage for everyone," he said.

"But I heard that you have long waits for some basic services in countries with universal health care."

"That's true. If you want an optional surgery like a knee or hip replacement, you're put on a waiting list and could wait six months to get your turn at the surgery. But when you have the surgery, you'll not incur any cost," Chuck said.

"In the United States, you could get in for the surgery much more quickly but it would cost a lot, even with insurance. That means that only the wealthy can afford the surgery. I guess in that case, the waiting time doesn't matter if you aren't wealthy. It basically means that you might never get a surgery that you need because you can't afford it," I said.

"I've also heard that people have lost everything, gone bankrupt, because of necessary medical bills. People need life-saving procedures to just survive and their insurance won't cover it or only cover a portion of it," Chuck said.

"Yes, I have friends that had to fundraise to have the money to cover their share of their medical costs. And that's from people that have medical insurance."

"Also, the costs are much higher in the United States," he said.

"I was diagnosed with Melanoma skin cancer in 2013," I said. "By the time they caught it, the cancer had penetrated my skin and gotten into my lymph system. I had a series of surgeries that removed most of the lymph nodes in my left leg at that time. Then I had a recurrence in 2015. Through that whole time, I was getting multiple CT scans, PET scans, and MRIs to monitor the cancer in my body. That combined with the surgeries ran up very costly medical bills. I had just started my job and had what I thought was good medical insurance. I remembered reading about a $2,000 cap on my out-of-pocket expenses that I thought was good. When the bills starting coming in, I found that the cap was true, but they had five different categories for surgery, hospital stay, tests, and other things. Each one had the $2,000 cap. So for three years, I was paying $10,000 a year in medical expenses. I was only making about $45,000 a year at that time, so it took almost all of my expendable income after housing and food. I definitely understand how medical expenses can bankrupt a person, even one with a good job and paying insurance."

"I've heard similar stories from Americans," Chuck said. "I saw a graph in an article lately that shows health expenditure graphed against life expectancy for over 20 countries. Every one of them has a higher life expectancy but a lower healthcare expenditure compared to the United States. The other countries spend half as much on health care and have on average a three-year longer life expectancy. The commonality between those other countries is that they have universal healthcare for their citizens. People get more preventative care and they get care for their medical conditions sooner when it's cheaper to treat."

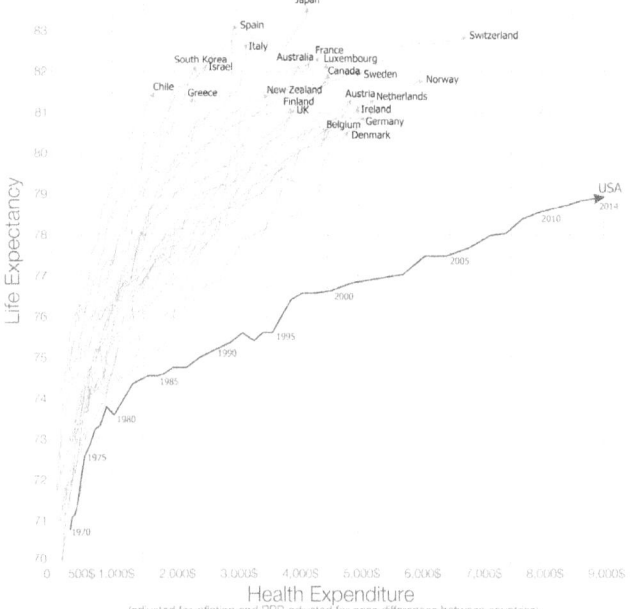

Life expectancy versus healthcare expenditures from 1970 to 2014. This shows that the United States spend much more on healthcare while not achieving as long of a life expectancy as other developed countries. This visualization is from https://ourworldindata.org/financing-healthcare/. [2]

"Yes. In the United States, many people will wait until they're very sick before they seek care. Much of that care happens in the emergency room when an ailment has been let go too long. Sometimes that's also the only place that people can get help when they don't have medical insurance. When it's a life threatening condition, then they have to receive attention."

"Also the administrative costs of your healthcare system are much higher than the costs in the other countries with universal healthcare. Most hospitals and doctors have a huge number of staff that deal with different insurance companies, filing claims, collecting bills. It becomes a major industry in its own right, just to collect money from people in your medical system."

"Even though you hear about the ease of getting medical treatment in the United States, it takes a long time to get an appointment for preventative care. Also we have a shortage of doctors that want to go into general practice because it doesn't pay as well as being a specialist."

"That does sound like a medical system that's not working very well," he said as he left with my blood samples.

I waited for about half an hour and the doctor came back in.

"Your bloodwork and EKG actually look good. I think you might have been having a panic attack. Have you been under much stress lately?" he asked.

"I've been working hard and been worrying about making ends meet with my income. So I could have been stressed from that."

"Well, take it easy and try to relax some. I suggest meditation or at least breathing exercises to relax. There are many useful free apps that will help you control your breathing and move you into a more relaxed state. I would recommend you try some of those. You're free to go."
"So, there is really no cost? You don't need a credit card or anything?" I asked a little incredulously.

"Nope, it's covered by the National Health Services. The most cost you will incur will be your taxi ride

to get back home."

I walked out towards the exit and the person by the front door asked if she could call a taxi for me. I thanked her and said that she could. A half hour later, I was back at my hotel at one in the morning. The entire four-hour ordeal only cost me thirty Euros for the taxi ride.

Chapter 8

Into the Mouth of the Beast

Robertsville, Indiana

June

Jenny and I made a rare trip to MallMart and parked in the massive parking lot that could hold more than six football fields. I had come to explore the new MallMart and try to understand the allure of this place for consumers and the motivations of the corporate entity. Walking up to the main doors I could feel the enormity of the space. This was truly a monument to capitalism where you could buy anything that you wanted at ridiculously low prices.

You could buy a bicycle for $70 and not even stop to think about how they produce a complex piece of machinery like a bicycle with gears and handbrakes for a mere seventy dollars. And it's even pre-assembled. That probably worth the $70 alone.

"When was the last time you were in a MallMart?" I asked Jenny. "I don't think I've been in one since that camping trip a couple years ago when we had to find fuel for the stove and it was the only 'camp store' nearby."

"I think that was my last time as well. It's always such a large place; I have always found it a bit overwhelming," she said.

"It's amazing the variety of items they sell here. You could buy parts to repair a car, buy toys for kids, shop for groceries, replace your entire wardrobe, eat lunch, and get your eyes checked without leaving this one building."

"That's part of the attraction, I think. You can get everything that you need in one place, which is great for a busy person or a single parent. It's a one-stop shop. I can understand the interest in shopping here. It just loses that charm when you buy another thing that breaks in just a week or you look at the labels of the items here and absolutely everything is made in China. It's not a very interesting shopping experience and the products are not made to last," she said as we walked past the toys and then the automotive parts over to the electronics section.

As we turned the corner we ran into Brady Ahlstrom. We had worked together at a small paper in Indianapolis for a short while. We were not very close, but we had gotten along well. He was tall and lanky and was wearing the red smock of the MallMart store.

"Hey Greg. How're you doing?" he asked.

"I'm doing well, Brady. It's good to see you. I don't think you've met my wife Jenny?"

"Nice to meet you," he said.

"When did you start working here?"

"About a year ago. The paper wasn't doing very well and couldn't keep me on as a writer. I was looking all over for work and this job came up. I figured I was over-qualified, but I wasn't getting any other offers so I applied for it," he said.

"I am actually writing an article on MallMart and how it fits into the community. Do you mind talking for a little bit?"

"Happy to," he said.

"How's it working here?" I asked.

"It's easy work. I'm a shift manger, so I just make sure everyone is doing what they should be doing to keep selling product. It's an interesting place to work, because you can actually advance and spend an entire career working here if you're a manager. My annual salary is $68,000 a year. I was surprised when I looked into it. The salary and benefits are good for managers. They are on par with the national average compared to other stores. If I continue and can become a store manager, I could earn between $150,000 and $250,000 a year with benefits.[1] It's not what I envisioned for my life, but it's a decent living."

"That sounds pretty good. I had no idea that a shift manager got paid so much."

"The cashiers and sales associates on the other hand, only get about $9 an hour," he said. "They don't start out with benefits and have to work more than 34 hours a week on average to get any benefits. Their hours are not very secure and of course we are open all the time except for Christmas. So some of the shifts are late nights and the hours can be long. Those entry-level positions don't have much opportunity for advancement."

"I've heard that MallMart is not very friendly with attempts to unionize."

Brady nervously looked around to see who was near enough to hear the conversation. "Don't say that too loud. They are opposed to that, and I could actually get fired if someone hears me talking about it. I have to check all of our employees in a database called UnionWatch to make sure that none of them are identified union members that have come in to organize workers. They take a very hard line on that subject."

"Do you know how they feel about the local community and other stores in the area?" I asked.

"Well, they don't support local sports teams like I see some of the local businesses doing. I have a son in soccer and their shirts never seem to have sponsors from Multinational Corporations. They keep a sharp eye on the competition and their prices. They occasionally have me go out while I'm working and 'shop' at our competitors, even the small mom-and-pop stores, to make sure that we are undercutting their prices. If I report that any store in town has an item that is cheaper than ours is, the store will actually drop the price on our item. It's a very cut-throat business," he said.

"How do they make a profit doing that?"

"They don't always make a profit on everything that they sell. They are willing to take a loss on a gallon of milk, for example, if it gets the customer into their store and draws customers from their competitors."

"That does sound pretty aggressive. You know that the Corner Grocer downtown that Nancy and George run?" I asked.

"Yeah, I remember that place. My mom used to send me there to buy stuff when I was growing up if she ran out of something in the kitchen."

"Well, they've closed. All of their business came over to the MallMart and they couldn't keep the doors open any longer."

"I'm sorry to hear that. They definitely were a nice couple and I used to enjoy my shopping trips there. I would get what my Mom wanted and they would usually give me a piece of candy," he said.

"Yeah, you don't get that type of connection anymore in these large stores. Well, thanks for the information. I'm glad to see that you're doing well," I said in parting. We shook hands and Jenny and I walked on.

"I never knew that you could have a full career at MallMart. What Brady has is a steady job where he can advance and he's definitely getting paid more than either of us," Jenny said.

"But he talked about how they are extremely competitive and will actively undercut anyone's price in town. They want to drive all other business out so that they can be the only option," I said.

"You have to admit that is a good business model. They have the capacity to dominate the market. Their economies of scale mean that they can operate on very low margins where they don't make much profit on any single item. Once they drive those other businesses out they can do what they want with prices. That competition is what keeps prices low," Jenny said.

"I understand the business sense of it, but isn't there a business ethic that needs to come along with this? I agree that a company has to make a profit just to be able to stay in business, create jobs for their employees, and provide a service to the community. But there is a limit to the amount of profit and how much destruction a company should be willing, or allowed, to do along the way," I said.

Chapter 9

Waste at City Hall

Robertsville, Indiana

July

The city of Robertsville was now our home. We had lived here for about a year and were settling into the community. Jenny and I were active and attended the city council meetings to keep up with what was happening. On this Thursday evening, they were having a meeting to decide the location of a new landfill that was needed for the town.

The old landfill was full to capacity and regulations stated that they could not put any more waste into it after the next year. It would be sealed off, made into a park, and used for methane capture that could be used as a heating source for a local brick making company.

When we entered the council chambers, we could tell that this was not a normal meeting. The council of seven members was seated along a panel-like table. They were all older white men that looked distinguished in their suits, but the room was packed with mostly African-

American members from the community. I remember seeing some of the audience members at a city clean-up event promoted by Keep Robertsville Beautiful. We were running late, so we got in just before the meeting started. We found a couple seats in the back and quickly sat down.

"I call the Robertsville City Council meeting to order," said James Carver who was the president of the council. "Our first piece of business is to vote on the location of the new landfill. It has been proposed that the site at Margaret and Oak is the best location of the new landfill because of its proximity to town to keep garbage transportation costs down, but it's also far enough out that any smell or blowing trash will not affect our community." He pointed to a large blown up map of the town and the location of the proposed landfill circled in red near the outskirts of the town.

A woman stood up from the crowd and shouted, "What do you mean it won't affect *our* community? I live three blocks from that site. There are whole neighborhoods that are within throwing distance of that site." With that the crowd erupted all talking at once.

James Carver seemed genuinely taken aback. "Order, Order!" he yelled and hit his gavel on the table.

After a minute, the shouting quieted down a bit.

"I'm afraid this in not the public comment period for this proposal. That was last week and public comment has been closed," he said.

"How'd you make the public comment period known?" another woman asked.

"We posted it in the city council newsletter as is required in our bylaws," James said.

"That's ridiculous. Who reads your newsletter?" she said.

"Since we are here today, sir, can you open up the floor for public comment?" another woman asked in a rational tone.

James looked around at the other council members who seemed to shrug or nod slightly. "I guess we can take your statements at this time. Please use the microphone and make your statements one at a time. These statements will be recorded in the official record of the meeting," he said.

The first person approached the microphone. She was an elderly black woman who was slightly stooped but confident in her walk. "Mr. Carver and fellow councilmen," she read from a prepared statement. "I come here today to make you aware of the egregious offense you propose to bring to our community. Your proposed location of the landfill is set to be placed directly in our neighborhood. I have lived in that area for 60 years. I grew up in that neighborhood. Most the people that you see before you today are my neighbors and friends. We come before you to say that the proposed site of this landfill in unjust. You place the burden of the trash from this town on our small community, a community that has been strong and coherent for generations. You propose to break that apart with your decision tonight to replace our community with a landfill." The crowd erupted with applause and affirmations.

"Order, Order!" Councilman Carver hammered his gavel on the table once more. "Thank you for your statement, but we must keep the responses to a minimum." "Dear Council Members," a young black man spoke into the mic next. "I believe that you move towards this decision, not out of malice, but out of ignorance. Our community on the outskirts of the town is close-knit and well established in this area. The last landfill that is now closing was also located in a high minority community 70 years ago. Back then we did not have the voice and power to protest this mistreatment, but today we will not be silent. You are taking part in environmental racism where your decisions from power burden minority communities. We demand justice." He returned to his seat to a round of applause from the audience.

These testimonials continued for some time.

After about ten others had said their piece, a white member from the audience came up to the microphone. "I support the statements of the community members that came before me and urge the council to rethink their location of the landfill. This is a deep-seated issue in our society that needs to be rectified," he said.

Next, a stately black woman with white streaks in her hair from age approached the microphone. "This is a clear example of discrimination against African-Americans. Environmental justice requires that you rethink your proposed location of the landfill so that you do not repeat the generations of injustice that have been perpetrated against our people," she said.

One of the people near us leaned over to her neighbor and said "That's Dr. Bennet. She's a professor of history at the university."

She continued with her statement. "If you map the locations of landfills and minority populations you will find a 74% correlation across the United States. Often time, white councilmen, because they are usually white and men, will make decisions to locate landfills in places that are not in their backyard. That decision process leads to locating landfills elsewhere in areas that the council members do not usually travel. This seems like a logical decision at the time but systematically discriminates against people of color. We will not let this stand."

Statements went on for about a half hour and the council members were talking amongst themselves and often nodding at the statements now.

Once the statements had been made and no one else was approaching the microphone, Mr. Carver said into the councilmembers microphone, "Do I hear a motion to withdraw the current plans for the location of the landfill to be further studied and relocated?"

"So moved," said one council member.

"Seconded," said another.

"All approved?" James said to the council.

With a chorus of affirmative votes, the motion was passed to remove the proposed site of the landfill. The audience members erupted in applause and many jumped up to hug each other.

"It's not often that you actually see democracy in action and the reigning powers actually listened to the public," I said to Jenny. "Do you mind if I stay on a little while longer and try to talk to James Carver?"

"That's fine. We're not in a rush," Jenny said.

After about 10 minutes as the room emptied, I was able to join Councilman Carver as he was walking towards the door.

"That was an interesting meeting," I said.

"Yeah, we really didn't expect there to be protest or even an audience at this meeting. I thought this was a done deal, but apparently we really did overlook a large part of the community and their concerns," he said.

"I was wondering if I could talk to you on another topic," I asked.

"Sure, what is it?"

"I was wondering about the decision process to bringing MallMart to town."

"Well, that was a few years ago. MallMart applied for tax abatement, building code variances, and tax increment financing for their project so that they could use their tax dollars to improve the roads in the local area around the store," he said warming to the topic. After this recent failure on the landfill, he was happy to talk about a project that the council did get done.

"So the city was willing to release MallMart from paying local taxes to bring the store to town. Why was that?"

"It was clear in their proposal that they would provide 40 jobs and an opportunity for the townspeople to buy food and items for cheaper. We knew the townspeople would want that."

"Well, definitely on the surface 40 jobs and cheaper products seems like a good opportunity, but loosing that tax base for improvements in the city can be pretty damaging. Especially with the increased tractor-trailer traffic that's bringing in products to the store. Also, did you notice all of the other businesses on the courtyard square that are going out of business? The Corner Grocer just put up a closed sign. The hardware store is in the process of closing and about three other businesses are closing up. We're losing the central business district and all of the purchasing power is moving out to MallMart, which is not even bringing in tax dollar revenue to the city. It's draining the vitality from the town and I'm not sure it'll come back," I said.

He looked downcast. "We hadn't considered those outcomes. Of course, I noticed Nancy and George's grocery closing up and the other shops down on the square, but I hadn't really connected it to MallMart coming to town. I just figured it was bad business planning."

"It would be an odd coincidence that the business plans for five separate businesses all failed within months of each other. I was talking to Nancy on the weekend, and they could see their clientele moving over to shop at

MallMart. The people of the town don't make enough money to have the luxury to shop anywhere for the benefit of the town. They choose the cheapest option just to survive because they don't have the disposable income to spend any other way. The inevitable effect is the loss of our small town businesses that made this town thrive."

"Well, I've gotten a lot to think about this evening. Thanks for coming to the meeting," he said as he departed with an obvious slump to his shoulders.

Chapter 10

The Local Food System

Robertsville, Indiana

July

As we were leaving the City Hall meeting, we ran into Joanne and Steven who were the grown children of Nancy and George of the local grocery store.

"Hey guys. I'm sorry to see your grocery store closing. I really liked that place," I said.

"Yeah, we hate to see it close as well," Steven said.

"But we have an idea about how to revitalize it and bring it back as something different that might be able to survive in the market with MallMart dominating the cheap grocery landscape," Joanne said.

"Really, what's that?" Jenny asked.

"We're organizing a start-up food cooperative," Steve said. "A cooperative business plan is where the community comes together and becomes member/owners of the cooperative by buying into the store. Their equity-shares provide the initial funds to get the store up and going. Members also invest in the store through a member loan program and then about half of the funds for the store are borrowed from a traditional bank."

"The only problem is that it's a long process to get the interest and backing of the local community members to invest in the store," Joanne said. "It's a much longer process than putting up collateral and having the bank finance a project, but it gets the community behind the store. We would focus on local and organic food, which the big box stores are not doing well. We would source many of our products, especially our produce, from local growers. That provides a stable market for them to sell their produce and provides good fresh food for our market."

"Nearly 80% of the funds from a coop stay in the local community compared to about 18% of the funds spent at a big box store that stay in the community through local jobs. The co-op will source as much food as possible locally, invest in the community, and provide a community service," Steven said.

"It provides local resources and helps to support the community, but it's not a not-for-profit. It's a business that means to make a profit, but has the support of the local community as one of its underlying missions," Joanne said.

"We're off to our first meeting of like-minded folks to get the co-op up and running. We plan to establish it in our parents' old store. We already own the lease on it and it has refrigeration and the HVAC that we need for a co-op grocery store. But we could use the influx of capital to give it some needed improvements. We can update the equipment and modernize the building. Then we will be a full-service grocery store with a smaller footprint, about 5,000 square foot retail instead of 185,000 square feet of retail for a standard MallMart store. Anyone can shop at our store, but member/owners get a discount. The member/owners also vote on the broad policy for how the store is to be managed and what we want to provide with the store. We will focus on local and organic, but the general manager will make the specific decisions about what will be sold in the store. We're excited about the project and can use our college training to bring these new opportunities to town," Steven said.

"That sounds great! How do I become a member/owner?" I asked.

"The standard member equity for many co-ops is $200 per household. This is a one-time fee and you are a member/owner for life," Joanne said.

"That's a lot of money to put into an idea that may not even come to pass, but count Jenny and me in," I said.

"Thanks! First we have to incorporate and decide on bylaws, but we'll come back and visit when we're ready to take memberships. We have the contact list of people who regularly shopped in the store, so we have a good starting point, but it'll take about 800 member/owners to conduct a viable capital campaign. But I think we can make it happen," Steven said.

"Great! I'll be looking to hear from you. I'd love to get our grocery store back and even better if it's focusing on local and organic and can keep more money in the community," I said in parting.

Chapter 11

Wealth Inequality

Santo Domingo, Dominican Republic
August

Jenny and I had been working hard at advancing our careers for many years and we just finished the move to our new house. We decided that it was time to actually take a vacation before our child was born. Also, my health scare in Ireland suggested that I should take the time to slow down and enjoy life a little bit. A vacation would help me relax. We wanted to do one more big adventure together before things changed with kids.

We decided to go to the Dominican Republic, which is on the tropical island of Hispaniola. We had heard good things about this country in the Caribbean. We were interested in the colonial history and the environment. Jenny got time off from work and we booked our flights.

We flew into Santo Domingo and got a shuttle to our hotel, the Renaissance Santo Domingo Jaragua Hotel & Casino. Santo Domingo is the capitol city of the Dominican Republic and the oldest permanent city established by Europeans in the western hemisphere. It has over 3 million inhabitants today and is a vibrant metropolitan area. It was the place that Christopher Columbus landed in 1492 and was later developed into the old city, some of which remains today.

As we drove up to the hotel, we were amazed at its opulence. It was a ten-story glass and cement structure that was surrounded by green grass and palm trees. The main foyer was high ceilinged with natural light streaming through the windows. It had shining marble floors and a sort of inverted pyramid-stepped structures accenting the ceiling. After checking in, we took the elevator up to the 9th floor and found our room down the hall. We had an expansive view out to the blue Caribbean Sea.

"This is amazing. I don't think I've been in a hotel this fancy in the United States, although I am sure they exist," I said.

"Yeah, and with the beach just outside the back door, this will be a nice place to hang out for a few days before we go to the mountains. I can't wait to explore the mountains and do some hiking," she said.

We settled in for the night and made plans to explore the city the next day.

We woke up early and had breakfast at the outdoor hotel restaurant with a wonderful view of the Caribbean. After that, we strolled down Paseo Presidente Billini, which wrapped along the coastline to the old city of Santo

Domingo, called the Ciudad Colonial. We walked down a large cement walkway that ran along the four-lane road with two lanes of traffic going in each direction. A row of palm trees separated us from the road, and the expanse of the blue Caribbean extended off to our right. We passed donkey-drawn four-wheeled buggies that were obviously there for tourists to get a leisurely tour of the town. We saw almost as many mopeds and bicycles as we did cars as we strolled along. A light breeze was coming off the ocean, which made it a pleasant 85-degree temperature with the sea breeze cooling everything off.

We crossed over the road into the old city where we could see the walls built in the late 1400s and early 1500s by the Spanish as they set up a vice regency on the island of Hispaniola. In 1992, this area was designated a World Heritage Site by the United Nations Educational, Scientific and Cultural Organization, called UNESCO for short.

Columbus first landed, inadvertently, on the north side of the island in what is now Haiti, as the Santa Maria went aground and sunk. Columbus set up a contingent of men there in a town they called La Navidad, but the town was destroyed by the time that they returned in 1493 and no survivors could be found. They decided to set up a new compound with 1300 men further east in what is today part of the Dominican Republic and called it La Isabela. A second town was established on the south side of the island called Nueva Isabela in 1496. A hurricane hit the town in 1502 and destroyed much of the area, so a yet another town was established on the west side of the

Omar River and called Santo Domingo. This is now the oldest continually occupied city in the Americas.

Christopher Columbus was, of course, not the first person to live in this area. The Native Americans, called the Taino, were well established in this area and estimates suggest that their population was over 1,000,000 strong at when Columbus landed on the island.[7, 8]

I read from a guidebook that the first smallpox outbreak among the New World inhabitants was recorded in 1518 and resulted in 90% mortality of the remaining native population that had not been killed in warfare or from enslavement. By 1548, only about 500 Taino still survived. A resurgence of quasi-indigenous Taino identity has occurred in rural areas of Cuba, the Dominican Republic, and Puerto Rico although much of their heritage was lost during early colonization.[6]

"You know, this is a classic social justice issue that we still are not facing in the Americas. We revere Christopher Columbus for finding the New World and establishing European inhabitants here, which includes our own heritage in the Americas. But we often ignore the native peoples that were here before," I said.

"It's the mentality of the victor. We bring warfare and conquer an area, so we believe that we have the right to own it. That's often the way that things happened in the past such as in the 1500s when much of this was going on. It's only today's awareness and sensitivities that give us the perspective that this might not be the way that things should be. Today, many people value the native cultures and their traditions and are trying to bring them back," Jenny said.

After walking the old city to explore for a while, we looked around for a lunch option prior to continuing our adventures. We walked by a Hard Rock Café.

"That's odd. A very American export right in Santo Domingo. It kind of stands out as brash and distracting. I'm not interested in that at all," I said.

"Look over there. That place called Kalenda. It looks like it fits in with the traditional surroundings much better. TripAdvisor says that it's Caribbean with vegetarian and vegan options. That sounds like some good local food. Let's try that place," Jenny said.

We had a lunch of Arroz con ropa vieja with fried plantains on the side.

"You know that strictly translates to rice with old clothes, but the roast beef is really good. They must stew it for hours. It just completely falls apart," I said.

"The beans and rice are really good, too. The flavors are exquisite," Jenny said. "And fried plantains are great. It's like a banana but a little starchier."

"I'm glad we stopped here. This seems like an authentic Caribbean restaurant," I said.

"It was great. Let's get an Uber over to the Three Eyes National Park," Jenny said as she typed it into the Uber App.

"I guess I shouldn't be surprised that Uber is here in the Dominican Republic. It really has taken over most places and is an economical way to get around," I said.

"It also gives you a chance to meet some of the locals and get suggestions and knowledge of a place. It allows you to experience the culture much more than a normal taxi ride or bus ride," she said.

"Yes, but the bus or public transportation options can often be a learning experience about the local culture and routines," I said.

"Yes that's true."

We waited outside for about 5 minutes and our Uber ride drove up.

"Hi, I'm Estabon. You are headed to Parque Nacional Los Tres Ojos or the Three Eyes? Hop in," he said.

"Yep, that's right," I said.

"Did you just eat at Kalenda? That's a great place that the locals eat at. I'm glad that you found it," Estabon said.

"It was really good. How long have you been driving for Uber?" I asked.

"About a year now. It's relatively new in Santo Domingo, but it's catching on. Many tourists come into town looking for it. They seem to be used to getting around by Uber," he said.

"Yeah, it's pretty popular in the bigger cities in the United States," I said.

"Are you going to get out of Santo Domingo and up into the mountains?" he asked.

"Yes, we are. We just need to figure out transportation," Jenny said.

"Well there is a bus that takes you out to a small town where you can rent a mule and guide to head up to Pico Duarte. It's a long bus ride though," he said.

"That's fine. We're not in a rush and it'll give us a chance to experience the country," I said.

"You can get a bus to La Cienaga at Terminal ASO-TRAPUSA in the center of Santo Domingo. Normally, it would take about three hours to drive there, but it might take about six hours on the bus," he said.

"It'll be an adventure," Jenny said.

"Good. Here we are at the National Park. Thanks for using Uber, and rate me if you want on the app. It helps others trust my services and increases my business," he said.

"No problem. Thanks for the information on the bus route," I said.

We stepped out of his car at the entrance to the Three Eyes National Park right in the city of Santo Domingo.

The Three Eyes were cenotes, which are natural depressions in the limestone bedrock that are fed by an underground stream. Interestingly for these, one of them had sulfur in the water, one had salt water, and a third was freshwater. The cenotes were some of the few places that had reliable fresh water resources in limestone landscapes. The surface water would erode the limestone, creating underground stream systems. But for people living on the surface, this meant that water was scarce and streams were rare. The underground caverns were extraordinary because they hosted their own unique ecosystems as well as providing freshwater.

Artifacts from the Taino culture were found in these caves indicating the native peoples used them. These types of cave systems were found throughout the limestone bedrock in the Americas. They were common in the Yucatan Peninsula as well as here in the Caribbean.

We paid our entrance fee and walked down the steep stairs into the cenote.

"Look at all of this rich green vegetation hanging down from the walls. The water is so dark! Look at the greens and blues! This's beautiful," Jenny exclaimed.

"It's amazing to think that the native peoples have been using these for thousands of years, and now it's one of the most popular tourist attractions in Santo Domingo," I said.

"It says in the guidebook that it was discovered in 1916 by Europeans and later developed into a park and recreation area," she said.

"These natural wonders are amazing," I said. "I wouldn't expect this in the middle of town. I can't imagine what it would be like to be the first person to stumble upon this. It was probably a Taino Indian a few thousand years ago that was out exploring and came across this cave. It would have been very hard to get into it without the modern stairs built into the wall."

"They would have had to lower themselves in by a rope or jump into the water and climb out by a rope. I wonder if this hanging vegetation could have held them?" Jenny said.

"I'm always amazed by the beauty of nature that we usually take for granted. These are astonishing areas, but we don't value them as highly as we should. We continue to pollute and use places like this for our own benefit. It's great that they set this up as a national park. That way, it is monitored and is safeguarded, but too many places in the world have not been protected like this," I said.

Chapter 12

The Lap of Luxury

Dominican Republic
August

We enjoyed Santo Domingo for two more days, exploring the interesting heritage of this town and trying the local food. The expensive hotels along the shorefront were much different than most of the city streets. We observed an obvious dichotomy between how most people lived in Santo Domingo and the rich tourist locations along the coast and in the center of town.

We took another Uber ride from our hotel to the central bus station with our bags and caught the bus out to Jarabacoa. The bus ride took six hours. Santo Domingo is a large sprawling city and a half hour passed driving by part of the city.

The New York Times put me in contact with Pedro Castillo in the area that had helped to establish Parque Nacional José Del Carmen Ramírez. I was able to get an invitation to his house to talk about the park. This park surrounded and protected Pico Duarte, which is the highest peak east of the Mississippi. The Mr. Castillo would direct us to La Cienaga where we would start our ascent of the mountain.

Once we left the city behind, our bus travelled up Hwy 1 through a meandering river valley that was now completely dedicated to agriculture. We were traveling towards the mountainous inner regions of Hispaniola, so most of the flat area was in the river bottoms. We passed small towns and villages stopping every now and then to let passengers off or pick them up.

I studied our fellow passengers as we travelled along. The bus was full, without any air conditioning besides the wind blowing in the windows, and smelled strongly of body odor. Three people sat in the two seats across the aisle from us; they had several chickens in their laps. It looked like they each only owned one set of clothes as what they were wearing was ragged and unkempt. They saw me looking at them and they smiled at me with infectious smiles. I nodded back at them and smiled myself. This felt like the raw Dominican Republic that you didn't see in the cities or along the tourist traps.

People were obviously travelling because they had to for commerce or to visit relatives. This was not a vacation or a luxury trip to them, but a necessity.

As I looked out the window, I could see we were on a two-lane road, but there were four cars abreast weaving in and out of each other to get a slight advantage of speed to pull ahead. Motorcycles and bicycles were coming towards us just along the side of our lane. Other motorcycles were going the other direction trying to beat the traffic.

One motorcycle had a family of five piled on it. The father was driving, the mother sitting behind him. She was leaning back so that a child could sit on her lap with another child in the father's lap. An older child was standing behind the woman on knobs that came out from the back tire axle as I used to have on my bicycle when I was younger. This child of about 10 years old on the back had one arm around his mother and the other holding a suitcase dangling off the back. The motorcycle was sagging under the weight of the family but making a valiant attempt to pass our bus.

The traffic was crazy. People were going every which way but they all seemed to be dancing to the same song. They sped past each other and made crazy turns but I didn't see any accidents. There were many narrow escapes, but everything seemed to flow to a pattern that was hard for me to discern. I could just imagine that a gringo like me trying to drive in this madness would stop everything because I didn't flow with the same rhythm that came naturally to the Dominicans.

"Look at the houses out here in the villages; they're so colorfully painted!" I pointed out as we rode along on the bus.

"There's an active baseball diamond in every town," Jenny said. "Didn't Sammy Sousa come from the Dominican Republic?" she asked.

"Yes, I think he did. Baseball is the main sport here and many young kids hope to pull themselves out of poverty through playing baseball. These villages are poor. They are a stark contrast to the opulence that we saw along the beachfront in Santo Domingo. There's a very large disparity between the wealthy and the poor here," I said.

"Hey, look at that! They're doing mountaintop removal here!" Jenny said sounding aghast and pointing out the nearby hills that were completely stripped of vegetation and a large portion of the top of the mountain was cut off.

"There's a sign at that entrance that says Falconbridge Dominicana Nickel Works. That must be a nickel mine," I observed.

We continued down the main road and saw passenger vehicles pulled over at the side of the road by what were obviously the local police. They had automatic weapons on their backs and wore uniforms consisting of dark grey pants with lighter grey button-down shirts. The cars said *Policia* on the side and had blue flashing lights on them. We passed a number of these stops on the side of the road and every police car varied. I even saw a Volkswagen bug police car. I gathered that their approach was not to outrun any culprits going down the road.

I leaned over to the trio sitting next to us and pointed out one of the police stops that we could see out the window. "Qué es?" I asked in my very limited Spanish for "what is it?"

"El soborno…payola…" he said rubbing his fingers together.

"Oh they are being stopped and have to offer a bribe to the police?" I said.

"That is an interesting money making scheme for the police officers," Jenny said. Apparently, buses were exempt from this treatment and we were allowed to pass without a problem.

As we drove on, we climbed in elevation into the mountains. We passed through tropical vegetation and were starting to get to higher elevation. There were a few hardwood stands in the canyon bottoms and West Indies Pine trees started to show up more frequently.

"I read about those," Jenny said. "Those are the southernmost naturally occurring pine trees in the world. Most pine trees occur in temperate areas in the north and aren't found at all in South America or Australia. So this is as far south as the genus *Pinus* extends."

"That's pretty cool. It's interesting to think that natural pine trees don't grow any further south than the Dominican Republic. We are so used to pine trees in the United States that you wouldn't think of them just ending without being represented south of the equator. That means the trees in the southern hemisphere must be very different," I said.

We arrived at Jarabacoa bus terminal, grabbed our bags, and walked into the station. Someone was holding a sign that said CUNNINGHAM in big letters. We walked over to the person.

"Hi, I'm Greg Cunningham and this's my wife Jenny. Are you here for us?" I asked.

"Yes, sir. Let me take your bags and follow me this way," he said in clear and efficient English.

He grabbed our two rolling carry-on bags and left us with our backpacks to follow him. He walked out to the adjacent parking area to a shining white Range Rover.

We hopped in the back seat and it felt like we were sitting above the driver with a wide view out the front of the window. The seats were leather, the windows were tinted, and there was a sunroof.

"This is a very nice car," I said to the driver.

"Yes, it just came of the assembly line. We get the latest model every year," he said.

I raised an eyebrow at Jenny. I was wondering what we got ourselves into.

We drove for about 30 minutes up a winding mountain road past fruit trees that were heavy with avocadoes, mangoes, and oranges.

We came around a bend and saw the expanse of a mansion in front of us with a scenic view into the valley. The viewscape showed the peak of Pico Duarte and lush forest climbing up the mountain. There was a stream down below but a manicured lawn surrounded the top of the hill where the house was situated. The size of the house and the scale and maintenance of the grounds made this look like a golf resort. We pulled up in the circle drive to a grand entrance to the house and got out of the car.

Pedro Castillo came down the steps to greet us. This was my contact. The New York Times had put me in touch with him since I wanted to talk about sustainability in the Dominican Republic while we were here on vacation. I had not known what to expect, but I would have never guessed that we would be visiting a billionaire deep in the woods in the Dominican Republic.

"Welcome! You must be Greg and Jenny. My secretary said that you had made plans to visit and would be arriving today so I sent my driver down to get you. How were your travels?" he asked.

"It was a great trip. We took the bus," I said feeling a bit awkward as I thought this was a person who had probably never ridden a bus and probably would not want to. "It was a great way to see the countryside and get to know a little bit more about the Dominican Republic."

"Come inside. I'll have the maid show you your bedroom. We can meet over dinner tonight in the main dining room," he said in parting as he handed us off to the maid to be taken upstairs.

We walked up a grand white marble staircase that curved up from the main floor to the upper levels. A glass chandelier dominated the center of the room. Large windows opened out on to the grassy expanse and the view of the mountains. All of the furniture looked like it was perfectly matched with the room and each setting looked like it was out of an interior design catalog.

We walked down a long hall on the second floor and the maid opened up a solid oak door to a suite of rooms that was as big as our house with a grand view out the window. We went back to the bedroom and dropped our bags and then came out to admire the living room and the view across the mountains.

"Gracias," I said.

"You're welcome. The main dining room is down the stairs and to the left. Mr. Castillo eats promptly at 8pm," she said.

"Thank you," Jenny said.

After the maid left, Jenny and I looked at each other.

"This place is amazing!" she said.

"I don't think that I've ever stayed in a place this nice in the US," I said. "It seems that our host is well off."

"He must be part of the very wealthy class here. There does seem to be a large dichotomy between the wealthy and the poor in the Dominican Republic," Jenny said.

"We don't pay much attention to it in the United States, but there is a strong wealth gap in the US as well. I just saw a video [3] that showed how Americans thought the wealth distribution looked and then how they thought it should be. Americans thought that the wealth distribution was skewed towards the super-rich but thought that it should be more balanced, that would actually provide a decent living for the all Americans. It turns out the reality is so far skewed that it's beyond the comprehension of most Americans. The wealthiest one percent of Americans control 40 percent of the wealth while the bottom 80 percent of earners share only 7 percent of the nation's wealth [3,4,5]," I said.

Rank and Person	Net Worth	Source of Wealth
1. Bill Gates	$81.0 Billion	Microsoft
2. Jeff Bezos	$67.0 Billion	Amazon.com
3. Warren Buffett	$65.5 Billion	Berkshire Hathaway
4. Mark Zuckerberg	$55.5 Billion	Facebook
5. Larry Ellison	$49.3 Billion	Oracle
6. Michael Bloomberg	$45.0 Billion	Bloomberg LP
7. Charles Koch	$42.0 Billion	Diversified
7. David Koch	$42.0 Billion	Diversified
9. Larry Page	$38.5 Billion	Google
10. Sergey Brin	$37.5 Billion	Google
11. Jim Walton	$35.6 Billion	Wal-Mart
12. S. Robson Walton	$35.5 Billion	Wal-Mart
13. Alice Walton	$35.4 Billion	Wal-Mart
14. Sheldon Adelson	$31.8 Billion	Casinos
15. Steve Ballmer	$27.5 Billion	Microsoft
16. Jacqueline Mars	$27.0 Billion	Candy
16. John Mars	$27.0 Billion	Candy
18. Phil Knight	$25.5 Billion	Nike
19. George Soros	$24.9 Billion	Hedge Funds
20. Michael Dell	$20.0 Billion	Dell Computers

These are the twenty wealthiest Americans along with their net worth and the source of their wealth. These are a few of the 1% that own the majority of wealth in the United States. This information is from Forbes in 2016. [4]

We unpacked our bags and cleaned up after our long travel. Then we took a walk around the grounds. The grounds were perfectly manicured and we saw gardeners that were working hard at maintaining this outdoor splendor. There was a formal garden in the back with trellises supporting wisteria and other flowering vines. All sorts of flowers were growing in beds and the whole garden was larger than two football fields. There were marble statues and fountains scattered throughout the garden paths that created unique nooks where you could sit and contemplate the beauty of the garden or the meaning of life. Bees and other pollinators where zooming around the garden doing their work. It was a beautiful setting.

We walked down the slope to the orchards where we strolled through rows of avocado and mango trees that were ripe with fruit. It was a lovely afternoon that we wished could last longer.

We returned to the mansion and prepared for dinner. We had not brought very fancy cloths so we did the best that we could. Jenny wore a dark blue dress that was a little more suited to the beach than a formal dinner and I wore some khaki pants and a purple button-down shirt for a splash of color.

We explored the house for a little bit while we waited for the dinner hour. We found an impressive library on the first floor. It was a large room that had wood paneling and arched gothic windows letting in natural light. All of the walls were lined with books in Spanish and English. Comfortable leather chairs were scattered throughout the room and an old wooden desk with a Tiffany lamp sat in the corner next to large windows. A rock-lined fireplace sat in the corner that looked like it was not used much, although it would create a nice atmosphere in the evening or morning when it was a little cooler.

We joined Mr. Castillo for dinner in the main dining room at 8pm.

"This is a beautiful place you have. We walked the grounds and really enjoyed your garden and the orchards," I said.

"Thank you. It has been in my family for generations. Today it makes money with the orchards. We have about 100 acres of orchard trees that go back amongst the hills. This valley has a unique climate on the island where it's high enough elevation that we get cool temperatures, but it's protected from frost. So we can grow a variety of fruit trees that can't be grown anywhere else on the island. It's not too moist so we don't have mildew issues either. We provide much of the tree fruit for the whole island. Most of what we grow is produce that is sold to the wealthy resorts on the north, east, and south sides of the island," he said.

"We have really enjoyed our stay so far. It's nice of you to accommodate us for an evening," Jenny said.

"You're very welcome here. I'm always happy to play the host. Anyway, my wife and children are in Europe right now exploring some of the archaeological ruins so I'm a bachelor for a few weeks and it's nice to have guests. What's the article that you are working on?" Mr. Castillo said.

"I'm writing an article that I'm calling Exposé on Sustainability. Sustainability is a vague term that means many different things to different people. If you ask someone what it means to them, most people in the United States would simply say recycling and most of the others would not have any idea what it means. But sustainability is a way of living that uses a minimal amount of resources to support our way of life, while preserving other resources for future generations. It involves living in balance with nature so that we don't use up the Earth's resources but we replenish them as we support ourselves," I said.

"Yes, many places around the world are thinking in terms of sustainability. Our own orchards that you saw are managed with sustainable principles. My family has been growing them for 60 years in this valley and you can imagine with our harvesting every year, we remove a lot of biomass in the form of nutrients being taken out of the soil and deposited in the fruit. We bring in natural fertilizers like seaweed and horse manure to replace the nutrients in the soil. We grow without pesticides or herbicides, but that takes a lot of manual labor. Luckily, we have a large and willing workforce in the surrounding villages. Many of the people in the surrounding areas have been working in the orchards for generations," he said.

"You must treat them well if they are able to support families and want to stay on for generations working in the orchards," I said.

"We do treat them well. We pay them more than they could earn with other jobs on the island, we take care of them when they're sick, and we even support them when they are too old to work in the fields. They appreciate the life that they have here so they are loyal to the orchards and my family. This was a concept that my grandfather started 60 years ago. That if we pay a good wage and provide what a family needs to live, grow, and thrive then we will have generations of workers. We get applicants from all over the island asking to come and work here, but we have the luxury of only taking the best of those applicants.

But once they work here, they know that they will be supported as long as they work hard," he said.

"That's also an important part of sustainability where the people are supported as well as the land. You are definitely displaying the basic tenants of sustainability when you can demonstrate generations of productive growth and generations of successful families. Why is it that you have decided not to use pesticides and herbicides?" I asked.

"I actually got my bachelor's degree from Yale and studied ecology and forestry. I learned about basic forest management that supports much of what my father and his father before him were already doing. Also from ecology, I learned how nature works in balance. That there is no waste in natural systems. The dead carcasses of animals and even their scat is broken down by insects, bacteria, and fungi to provide nutrients in the soil for the next generation of plants. I could see that as we remove fruit or any biological material from the orchards, we take away nutrients. So those nutrients need to be balanced by the input of other organic matter. Because of this, we leave branches and leaves in the orchards and bring in other organic fertilizers to balance what we take away. The insects, bacteria, and fungi are necessary to maintain the cycle of nutrients so we don't use pesticides because it would damage that cycle," he said.

"This concept of balance is an important part of sustainability as is the long-term view of supporting the environment and the social systems that your company depends upon. The other part of sustainability is the economic side where a business must be able to support itself to be sustainable. So a company needs to be able to make a profit," I said.

"Yes, we do well in that area as well. Because we can market our produce as organic and it's the only source for these fruits on the island, we can set a high price. The resorts are willing to buy at that price and they get the benefits of selling organic and local products to their clientele. It makes the cost and the effort worthwhile," he said.

"I can't help but notice that you live quite well. I imagine that your income from the operation far exceeds the average worker," I said.

"That's true. I probably make 600 times as much as my average worker does. Labor is cheap in the Dominican Republic and as I mentioned we can choose the best candidates because we offer good working conditions. Because the expectations of workers are so low, we can offer a little bit more and be a good opportunity for them. We deserve the higher income that we receive because we have planned and engineered this opportunity for all of us. My family has worked for generations to grow this successful business and we are now operating with economies of scale where the individual fruit that we produce do not cost us that much in inputs, but we can sell them at a premium," he said.

"But most of the homes that we drove by throughout the Dominican Republic and as we came up the mountain consisted of dirt floors with tin roofs. They had very basic living accommodations while your estate is palatial. Don't you think that some more equity would improve everyone's situation even more?" Jenny asked.

"Listen, you have to consider where the workers come from. They don't expect more than a dirt floor, tin roof, and a little bit of electricity from a solar panel. That's all they have ever known in their lives and it's what they expect. We make sure that they are doing better than they could have in the villages that they originally came from and that's enough to keep them content and loyal to the company. That's the economic sweet spot that we try for and everyone wins," he said.

"We definitely don't mean to insult your hospitality and it seems that you have built a sustainable business where everyone, including the environment, benefit. I applaud you and your family for your efforts. What I was really hoping to talk about is the establishment of the Parque Nacional José Del Carmen Ramírez and the adjacent Parque Nacional José Armando Bermúdez . I understand that your family was involved in the establishment of the parks in the 1950s," I said changing the subject.

"Yes, my grandfather originally purchased this land at the base of the mountains in the 1940s. He had some money, but was not very rich. He bought a little land and started the orchards. As they did well, the trees matured, and production increased he acquired the funds to buy more land. By the mid 1950s, more people were coming into this region and were cutting trees from the forest to build houses. My grandfather saw that the environment was being damaged. Heavy rains would wash away the soils and the sediment was building up in our valleys," he said.

"That was rather forward thinking of your grandfather to put together the logging in the mountains with local environmental effects," I said.

"Yes, he was always a visionary and could put together big picture ideas. He started to lobby the government to set aside the highest mountains as a national park. Pico Duarte has always been a sacred place to the Dominicans. Our people make a pilgrimage to the peak at least once in their life. The mountain is the dominant feature on the island. It is 3,096 meters tall, which is just over 10,000 feet I believe. It controls our climate, catches our rainfall, and drives our river flow. It means life for this island," he said.

"But the establishment of a national park also took away a resource that the people were using up to that point," I said.

"Yes, nothing happens without struggle. People were illegally cutting trees in the mountains to sell for timber or to build houses. There was no control on the use of the land. So the government put it under protection by designating it a national park in 1956 for one of the parks and 1958 for the other, which protects most of the mountain now. But the government also did a wise thing. It was a loss of resources for the local people that were using it, but the government made a rule that you must have a guide and a mule to take you up the mountain. This started a new ecotourism industry surrounding the park that didn't exist before. Now the village of La Cienaga has a stable source of income taking pilgrims up the mountain," he said.

"That's an interesting solution: the locals can obtain a livelihood from the protected area so they get some benefit and have a stake in the protection of that area," I said.

"Yes, but there's still illegal logging that occurs and illegal fires are set to promote some types of vegetation or for hunting. We have a few guards that patrol the mountains, but it's such a large place that it's hard to manage," he said.

"How stable is the protection and management of the national parks?" I asked.

"Sadly, it varies with each administration and political party that comes into power. Our political system is volatile which results in a lack of stability through different administrations. When a new party comes into power, it might decide not to maintain the meteorological stations so we don't even have good climate measurements through time. Protection of the national parks also varies with administrations, which makes their long-term survival precarious. But even with that political uncertainty, we have about 80 percent voter turnout," he said.

"That compares to a 72 percent turnout in England for Brexit and 58 percent turnout in the US in the 2016 presidential elections. So Dominicans are more active in politics than we are in the US," I said.

"Those are interesting statistics considering the US is seen as the leader of democracy. Still, I'm concerned for the future of our national parks with the volatility of our administrations," he said.

"We are looking forward to our trip there tomorrow. I can't wait to see the landscape and ride the trails," Jenny said.

"I'll bid you goodnight. I need to do some work in my study. You are free to use the house and make yourself at home. *Mi casa es su casa.* My driver will take you to La Cienega at 6am tomorrow morning as the guides like to leave early. I have also sent up a man to make arrangements so you have your trip reserved for tomorrow," he said.

"Thank you so much for your hospitality and the information about your operations and the national parks," I said.

We had an after dinner drink in the library where we spent a pleasant hour reading through books that we found interesting on the shelves and then went to bed in our elegant suite.

Chapter 13

The Mountain

Dominican Republic
August

The next morning, we were up before sunrise, washing and packing. We came downstairs to an elegant breakfast of fresh fruit, hot porridge, and fruit juices along with fresh baked breakfast rolls and breads.

We met the driver at 6am and headed to La Cienaga.

It took about an hour to get there on an older, partially paved road. It looked like the road had been paved at some point in the past, but floods and rainstorms had washed away much of the paving, leaving more potholes than asphalt.

We drove into a small town situated next to a river where we could see pens of mules near a central building that was larger than the others. Our driver parked and went over to talk to some of the men that were sitting around the house. It looked like arrangements were being made on our behalf.

We quickly repacked our possessions so that we just had what we needed in our backpacks for the two-day excursion. We brought along one change of clothes, ground pads, and compact sleeping bags. The weather was not supposed to be too cold, although we were going to higher elevation so we packed as light as possible from what we brought in our luggage to the Dominican Republic.

The driver came back over and pointed out two guides that were headed over to get a couple of mules. He said that we were all set and that he would pick us up in two days at this same location.

"Can you talk to the guides about storing our suitcases here while we take our backpacks up the mountain?" I asked.

"Why don't I just take them back to Mr. Castillo's house and I'll bring them with me when I come to get you," he said.

"That would be great. Thank you!" Jenny said.

We left our bags in his capable hands and headed over to the guides that were saddling the mules. The mules looked small to me. I'm not a very large person, but the mules and saddles were definitely made for Dominicans that are generally smaller than I am. The guides did not seem too concerned, so I trusted in their judgment that we would not break the animals.

We paid the lead guide for our four-day trip with three mules and two guides, which cost $11,547 Dominican Pesos, which is worth just over $250 US dollars. Not bad for a four day trip in the mountains with three mules and two guides; it was also a good income for the local guides.

After a few minutes of tightening girth straps, the guides motioned for us to give them our packs and to climb up on the mules. This was a new experience for us, but we gamely did as we were told.

Our packs were put into panniers on a third mule and we mounted two others. Our two guides seemed content to be walking.

We headed up the stream valley and soon started to climb up the mountain. We passed through a mahogany forest of hardwood and broad-leaved trees. The sun came dappling through the canopy to speckle the ground. A soft breeze was blowing, making us glad that we were wearing our fleece jackets. The mule ride wasn't bad; they just plodded along gradually making distance. I felt rather like a skilled cowboy as we maneuvered the stream and meandered down the trail.

After about an hour of going up the river valley, we turned onto a mountain path that steeply climbed out of the valley. Riding on an incline strained my riding skills as gravity was trying to pull me of the back of the mule. I had to use all of my muscles to hang on to the mule.

"Look at those views out over the valley," I said to Jenny as we travelled along and I pointed out a view over the mountains.

"This is really gorgeous. It's so wild and raw compared to most places I've been in the United States. The trails aren't as maintained and wide and we haven't seen anyone else on the trail. It just feels like we are in a more private place than the parks in the US," Jenny said.

We passed through different vegetation zones. We left the dense hardwood forests of the stream channels and the West Indies Pine was becoming more common. At around noon, we stopped at an overlook with a few benches. The guides tied the mules to metal railings that seemed to be installed for this purpose. As I tried to get off the mule, I realized just how hard my body had been working on the ride up the mountain. I nearly fell over stepping off the mule's stirrups, and I walked around stiff and uncomfortable.

"I thought that the mule ride would make this ascent nice and easy, but I can hardly move now," I said.

"We don't normally ride mules, so we are using a whole series of different muscles," Jenny said although I couldn't help noticing that she was moving a lot better than I was.

We got cheese, sausage, and water out of our backpacks in the panniers. We walked around the lunch spot; it was a flat landing in a little saddle in the mountain. Most of the rest of the terrain that we had come through and were surrounded by was steeply sloped. We rested for 30 minutes and then reluctantly got back on the mules. The guides seemed to be in great shape. I'm sure that they do this all the time, but they were happy to be hiking along and seemed content to move at the pace of the mules and didn't tire.

As we continued up the trail, I noticed that the ground level kept getting higher.

"Jenny, check out the erosion on the trail. Ground level is now at about head height and we are sitting on top of mules. That must be about 10 feet of incision on the trails," I said.

Jenny reached out and scooped some of the rock wall out by hand. "You know, this looks like granite, but it's completely weathered and breaks apart easily. I imagine the warm temperatures and heavy rain has caused fast chemical erosion so that the granite bedrock is mostly weathered in place. Then when the rains come, the trail acts like a river. The energy of the rain must shoot down the mountain and carve the trail deeper."

"That could be why we have the logs crossing trails in the US. They divert the water off the trails and keeps them from eroding like this one has," I said.

We travelled on for the rest of the day with the slow plod of our mules. We reached the top of Pico Duarte about an hour before sunset. The mules came up to a wooden structure and the guides started to unpack everything. They gestured for us to take our things inside where we found a large open room that provided protection from the elements. Above the door was a sign saying *Refugio*. We set our ground pads on the wood floor and spread out our sleeping bags. These were nice accommodations for a hiking trail.

"Want to go and explore the peak? We could probably get up there for the sunset," I said.

"That sounds great. I can carry some water and food in my pack now that most the stuff is out of it," she said.

We headed up the trail from the little compound and in about ten minutes came out to the opening at the top of Pico Duarte. We could see all around us as the mountains dropped off in every direction. The forest now was all West Indies Pine and just below the peak was a field of large boulders, which we saw on the map was called *Conuco del Diablo*.

"Before we came, I read a paper on a climatic reconstruction from this very spot. They were able to use tree rings from here to compare the tree growth to climate data from Jarabacoa and Constanza on the other side of the mountains. They showed that these very trees recorded temperature for the past 300 years [9]," I said.

"It sure is nice to get out from the bustle of our daily lives and enjoy nature," Jenny said as the sunset over the mountain peaks was painted yellow and red from the slanting sun.

We stayed on the mountaintop for a day and explored the surrounding area. We could see where landslides had occurred from the recent hurricanes that hit the area: Hurricane George 1998, Dean 2007, Irma and Maria both grazed the country in 2017. One mountainside showed 150 landslide scarps that we could count from where we were sitting on the peak.

Our guides saddled up the mules and we headed back down the mountain the next day. It had been nice not to get on a mule the day before. I realized that I had saddle sores where the constant rubbing in the saddle had irritated my skin and left sore scabs. Now I could feel those even worse as I climbed back into the saddle.

We headed down the mountain and retraced our steps through the eroded trails. Once we got about half way down, the mules knew that they were approaching home. They picked up the pace until finally they were running down the mountain. We pulled back on the reins but it did not do any good. At one point my mule took a wrong turn and got off the trail. It continued to run and came up to the cliff edge that marked the edge of the trail where it stopped abruptly throwing me off the mule to tumble down the hill onto the trail.

The guide came up and helped me up. I didn't seem to have anything broken, but I was bruised in several places and it felt like I had dislocated a shoulder. After, that, I decided to walk the rest of the way. I did learn the meaning of the phrase running for the barn and that will stay with me forever. The mules were definitely ready to get back home to their barn and I had never seen them move so fast.

When we arrived back in La Cienaga a little more battered and bruised, we expected to see the Range Rover waiting for us, but no one was there. We sat by the mule shed and waited for a few hours. As night was coming on, I went over to the head guide that we had ridden with and asked in my halting Spanish about Mr. Castillo's car coming to pick us up.

The guide went off to talk to a couple of people and came back shaking his head. He said that no car had arrived. Night was coming on and we must have looked worried, because he went off to talk to a couple other people.

A younger guide came up that we had not met before.

"Hello, I'm Pablo. You missing Mr. Castillo?" he asked in broken English, but enough to understand.

"Yes, we we're supposed to meet him here today," I said.

"Come stay at my house tonight. Mr. Castillo maybe come tomorrow," he said.

"That's really nice. We'll be happy to stay with you," I said.

We put on our packs and walked down the road for half a mile and stopped in one of the wooden sheds that we had seen on the way up with the tin metal corrugated roof. The building consisted of two rooms that had dirt floors. The back room must have been the bedroom and the rest of the house was the front room where everything else happened. There was a fire pit out back, which is where they cooked, and there was a communal toilet that was a pit in the ground with a slight wall of tree limbs around it.

Pablo went off to talk to his wife.

He came back with five kids trailing along behind him in a high-energy pack and his wife came wiping her hands on a towel.

"This is my family. They are happy for you to stay the night. This is my wife Camila. She will prepare dinner for us," Pablo said.

"Thank you. You're very kind," Jenny said.

"Come with me. You can leave your things here," Pablo said.

We went out the front door and the two oldest children came with us. We headed down to the center of the small town. We could see lights coming on around town. I hadn't noticed it before, but I could see small 1-foot by 2-foot solar panels on most of the houses that connected to a car-sized battery. This seemed to be what was running about one light per house.

We stepped into a slightly larger building that had signs advertising soda and beer on the outside; it turned out to be the local cantina. Chairs were scattered around the perimeter of the room. The floor was cement and there was a small bar in the corner where Pablo ordered five beers. The young kids that must have been teenagers got a beer, as did each of us.

"Thanks. Is this a special occasion?" I asked.

"No, this goes on every night," Pablo said.

Someone turned on a radio. Sounds of salsa, merengue, and samba came through the night air and couples danced in the middle of the small space. It seemed that everyone knew each other and partners often changed. Everyone was smiling and laughter was the only thing heard above the music. Everyone was interested in us as we were the only unusual things in the room.

Pablo leaned over to us, "You are the only gringos to stay in the village for years. People are quite interested in you."

"It's an interesting experience to be a minority. Where I'm from, we're part of the majority and I haven't experienced being the unusual one in the room," I said.

"It changes your perspective when you're the only one of your kind in an area. It's hard to blend in. You naturally stand out as something different," Jenny said. "Come on, Greg, let's dance."

We moved into the crowd on the dance floor and watching the others steps and worked at learning the salsa. Everyone around us seemed to flow with the rhythm where we definitely were working to try to keep up.

People laughed good-naturedly at our attempts and finally a couple broke up and came over to show Jenny and me how to dance the salsa. By the time that the song finished, I felt like we had improved our dance moves.

We stayed for another hour. I found out that I liked the merengue because it was a simple step, rocking back and forth. Pablo motioned to us and his kids that it was time to go. We said goodbye and thanked our dance instructors and headed back out into the night.

The roads were dark but we could see single lights burning in a few homes, most likely powered by the solar panels and batteries that we saw before.

"That is the happiest group of people I've seen in a long time," Jenny said.

"Yes, once the day's work is done, we come together and dance and enjoy each other's company. It's a simple thing but it's what we live for," Pablo said.

We came back to his home and his wife was serving chicken and rice on mismatched plates.

"Thank you," Jenny said.

I gladly accepted the plate and started to each the chicken. "This is amazing. The meat just falls off the bones and it tastes so good. Where is this from?" I asked.

Pablo pointed to the backyard. "Camila and the kids just killed it, plucked it, and boiled it on the fire out back. We keep chickens for eggs and meat on special occasions."

"This is the best chicken I've ever had. It must be because it's so fresh. It tastes very good. Thank you for your hospitality," I said.

The evening passed in a pleasant way. I offered to pay Pablo for the food and lodging, but he declined. "It's not necessary. I'm happy to help you. You will be talked about in the town for months to come," he said with a smile.

We settled our sleeping pads and bags in the front room and drifted off to sleep. Between the mule rides, hiking, and excitement of the day, we both fell into a heavy sleep immediately.

The next morning I was just waking up with the sun coming in when I heard a honking horn outside.

Pablo came in from the front door. It looked like he had been awake for a while.

"Mr. Castillo's driver is here for you," he said.

We quickly packed our few items and headed out front.

"Sorry that I missed you yesterday. Mr. Castillo was called away and had to go to the airport so I was unavailable yesterday," the driver said.

"No worries. It was a great experience to spend the night in the village and get to experience a little bit of life in the Dominican Republic," I said.

"I will take you to Jarabacoa where you can catch the bus back to Santo Domingo," he said.

We thanked Pablo and offered to pay him again which he declined. We waived goodbye and got into the Range Rover. With our five-day adventure in the wilds of the Dominican Republic, we felt quite a bit dirtier than the last time we were in the Range Rover.

The driver dropped us off at the bus terminal.

"Thank you for taking the time to drive us around. Please thank Mr. Castillo for his hospitality," I said.

"You're welcome. Mr. Castillo asked that I pass along his good wishes. Have a good trip home," he said and then drove off.

We got tickets for the bus, which was headed back to Santo Domingo in three hours. When we boarded the bus, we felt like we fit in a little bit better and had a better understanding of the culture.

Chapter 14

Walkable, Bikeable Cities

Terre Haute, Indiana
September

Green Town was having a conference in Terre Haute, Indiana, which is located just off I-70 near the Indiana and Illinois boarder. I decided to drive over for the day and catch this sustainability conference.

I was leafing through the brochure which had tracks called Food and Healthy Living, Community Sustainability Planning, The Built Environment, Living Sustainably, and Our Energy Future. As I walked into University Hall, I ran into Pat Marvin. He was the city planner for Terre Haute and worked hard to get funding to build biking and walking trails, and to clean up brownfield sites.

"Hello Pat. How're you doing?" I asked.

"Good. Busy as usual, but it's nice to take a day and talk about sustainability with like-minded people. Too often, I'm hitting my head up against brick walls in the city offices," he said.

"Oh, what're you working on now?" I asked as we walked through the newly remodeled halls and found a break-out room dedicated to Community Sustainability Planning.

"I'm trying to clean up a brownfield site on the east side of town. We have an Environmental Protection Agency grant and the site is badly contaminated with dry cleaning solvent. It qualifies for assistance under our grant because the company that dumped the chemicals on the site is no longer in business. But the property is right next to some city property and the city doesn't want us investigating the site in case it leads to an expensive clean-up that they will be stuck holding the bill for," he said.

"So it comes down to money rather than doing the right thing," I said.

"We have grant funds to help clean it up, but we need to get the survey done and start the work before the grant expires. Otherwise, anyone could sue the city to do the same clean-up without the benefit of the grant money. The city doesn't want the trouble and the perception of the contaminated site getting out in the public, but that is likely whether or not we get to work on it. It's just so frustrating that the city won't be proactive and take care of the problem. They just want to bury it and hope it goes away," he said.

"At least you don't have any Superfund site in Terre Haute," I said.

"No the closest superfund site is in Monroe County around Bloomington. The closest Resource Conservation and Recovery Act site called RCRA and pronounced like rickra, is about that far away as well. The Superfund sites are bad cleanup sites. They usually take millions of dollars and are supported by a tax on polluting companies. RCRA sites are usually landfills or hazardous waste dumpsites. Brownfields are bad but not nearly as bad as RCRA or Superfund sites. They are usually smaller with one particular problem because of a company that was dumping illegal chemicals onto the ground at one specific site," he said.

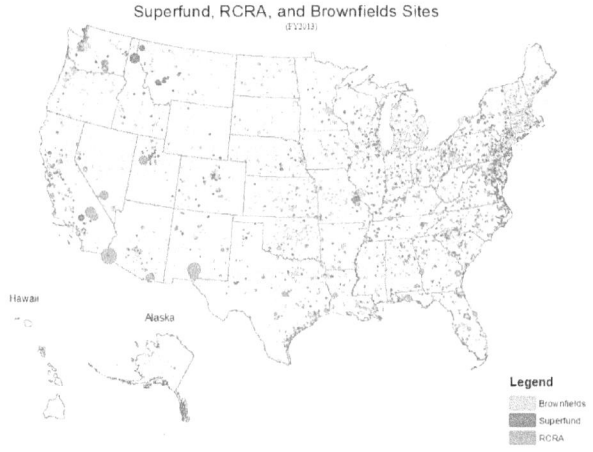

Superfund, RCRA, and Brownfields Sites

A map of the superfund, RCRA, and brownfield sites throughout the United States. This information is from the Environmental Protection Agency. [1]

"Still they all sound like a serious mess," I said.

"Yeah, they all take a lot of effort to clean up and can be very hazardous for groundwater, wells, and drinking water," he said.

The session started and we heard about how South Bend, Indiana had invested in its downtown area. They made the streets walkable, entrepreneurs bought up old, abandoned, industrial buildings and converted them to high-rise housing, and businesses started coming back to the downtown area. Now the city center is a place that people want to be. It was always bustling : during the day, with business people and at night with restaurants and bars. All of the increased activity improved revenues and provided a surplus to the city that they could continue to work with to improve the now vibrant city.

The next speaker talked about how South Bend had started with a Complete Streets policy that required that any new road construction consider walking and bicycling accessibility. This was an easy policy to implement and did not require much regulatory control. But it made everyone in the city, from the planners to the construction workers, know that biking and walking should also be considered along with cars when road construction is changing the roadways. The end result was an extensive connected trail system in South Bend. People could walk and bike throughout the city and that resulted in a healthier population.

The final speaker in the set was Samuel Lofgren who spoke about global Blue Zones, areas in the world that had the oldest living individuals and the talk also focused on what actually brings happiness. It turns out that the things that drive blue zones are a good diet, often with fish, and daily exercise, usually through effort to obtain food. The researchers expanded on these ideas and examined US cities to see which where the happiest and found a number of factors that controlled happiness. Part of it was having a good environment to live in, but much of it was a combination of being active with walking and biking, access to community and friends, and the ability to make a difference with their work [2a].

At the end of that session, there was an open workshop to do a walking inventory of Terre Haute and examine the town's walk score with Dan Bruin [2].

We met by the fountain in the middle of campus and Bruin started to talk to the crowd of about 50 people. "The walk score is based on a number of factors that look at how walkable a city or community is. This depends upon the businesses to which you can walk. Can you walk to get fresh groceries? Can you walk to the bank? Can you walk to restaurants? What is the quality of the sidewalks?" Dan said.

"Who developed the walk score?" I asked from the crowd.

"It was developed by real-estate brokers. They recognized that it was more valuable to be in a walking community where you could easily walk and get all of the resources that you needed, so they developed this score to help promote areas and increase sales of houses and apartments," he said.

"Let's go for a walk!" he said as we left the fountain and headed over to 3rd street, which is the busiest street in town.

"Now look here, you have sidewalks and walkways that don't line up. The main design is made for car access which cuts right across the pedestrian path and there is not even a crosswalk or connecting walkable path for pedestrians," he started demonstrating the failing of the town's walking infrastructure within 50 feet of leaving the fountain.

"I'm afraid that Terre Haute is not going to do very well with a walk score. Not many people walk here and it's likely due to not having good access. Since it's a college town, you would expect a lot more people walking the streets and more food and drink places in the downtown area," I said to Pat.

"Yeah, we have 22,000 students between four institutions of higher education and the town itself is only 60,000 people. You would think that the student population would be a large economic driver and that companies and the city would cater more to making Terre Haute a place they would want to hang out. But it's just now getting to the point where some people are starting up that kind of business in the downtown area," Pat said.

We walked over to 3rd street. The organizers had decided that the street with three lanes of traffic in either direction, was too busy and not well signed enough to try to take 50 people across on a walking tour. They turned south at the road.

"Now this is the narrowest sidewalk I've ever seen. It's too wide to be a curb and too narrow to be a sidewalk. It measures just a foot and a half wide. Why any city engineer would put something like this in is beyond me," he said.

"It does look like an afterthought or that someone cut off part of it in building the road," I said.

"Don't look at me," Pat, the city engineer said. "This was like this when I moved to town 30 years ago. I've no idea what they were thinking."

"One solution for this road and many others is a lane diet. If you reduce the traffic lanes to 10 feet wide, the traffic automatically slows. The cars are closer together so people are aware of that and slow down for safety. That also frees up space for pedestrian walkways or bike lanes. You can also put planters in the middle of the roads with green space and low vegetation. This also causes the traffic to slow and provides a safe haven for pedestrians if they are crossing the road," he said.

"Indiana State University did that on 7th street. We reduced two lanes of traffic in each direction to one lane in each direction and a center vegetated island. We increased lighting and made a much nicer roadway along one side of the main campus. It really helped to slow traffic and give more space to the students. It's also one place where we put in bike lanes," Pat said to me.

"One thing that I see missing in this whole area are bike lanes. It would be hard for a bicyclist to get to anywhere important in this city. Improving the trail and biking infrastructure would greatly improve access and reduce traffic in town," Dan said.

We continued our circuit down Wabash Avenue and through the downtown. Recent city work had put in curb cuts so that handicapped and bicycles could at least cross the road without having to jump a curb.

I was learning a lot about how to make a city more sustainable. It seems that just walking and biking access along with downtown housing and eyes on the street made the downtown area safer at night. Of course, good lighting in walkways was important for safety, but all of these infrastructure changes could improve the connectivity and vibrancy of a city. Who wouldn't want to live in town where you could park your car and spend the rest of the day walking to everything that you needed to get to, or better yet, live downtown and not even own a car?

As we were headed back up to campus, a swarthy young man stepped up and started a conversation.

"That was a good tour. I understand the importance of walkability and bikeability much better now," he said.

"Yeah, me too. Although I see that Terre Haute has a lot to improve before it can achieve that vibrant downtown feel that South Bend has been able to do," I said referring to the talk sessions that I saw earlier in the day.

"It sounds like you know a lot about sustainability. What do you do for a living?" he asked.

"I'm a journalist working on a freelance article for the New York Times on sustainability," I said. "What do you do?"

"I happen to be a sustainability professional. I run a small sustainability office in the town of Jasper. Maybe, I could provide some information on sustainability for your article," he said.

"That would be great?" I said a little vaguely. I had already set up many contacts and had a good direction. I was always happy to take on more information, but this offer seemed a bit odd.

"What do you have for your article so far and what direction are you taking with it?" he asked.

"I'm just gathering background information at this point and taking notes at conferences like this," again my instincts were telling me to be vague. Something wasn't right about this interaction. "I don't think that we have formally met. What's your name?"

"I am James Bedford," he said but didn't ask my name.

Luckily we were getting back to campus so I was able to excuse myself and start heading off in a different direction.

"I'll be in touch and maybe we could meet in Indianapolis to talk about sustainability some time," he said in parting.

I just waved and walked. I was starting to feel wary of this interaction, since I never told him that I lived in Indianapolis and he never asked my name. I would have to keep an eye out for him in the future and see what he was up to.

Chapter 15

The Wedding

The Farm near Clay City, Indiana
End of September

It was a warm September day. It felt like summer was just not going to end, but it was a nice day for an evening wedding. The wedding planning went quickly and Jane and Jessica put it all together with a little input from Chris, although I think he was mostly happy to leave it to them. They rented a large tent for the reception that could easily hold 100 people out behind the house on the grassy lawn. They had set up rows of chairs down by the lake with a decorative arch framed by two large sycamore trees for the ceremony. It was a nice, simple, country wedding. The pastor from the church that Chris and Jane attended presided over the ceremony.

Jane had picked flowers from the farm and put them in bouquets along the aisle. Her own bouquet was a beautiful set of Echinacea, black-eyed Susans, and Zinnia. She wore a white dress that was her mother's wedding dress and Chris wore his formal Sunday best suit.

They had each chosen passages from the Bible that were read by relatives and friends during the ceremony that talked about growing and nurturing for their coming life together.

They had hired a local string quartet to play music during the ceremony.

The ceremony started at 7pm and there was a light breeze, cooling off the heat of the day. Jane was beautiful with her long blond hair done up. She was just a little bit shorter than Chris with a slight build, although she was fit. She had worked on a farm all her life, so she had the strength of a farmer.

All of Jenny's relatives were there. Her two uncles with their families and her grandmother were in the front row. Jane's family was also out in large numbers. She had grown up only 10 miles from this farm, so she also had a lot of family in the area.

Jenny was a bride's maid and I was a groomsman.

It was a beautiful setting and I had never seen the farm so nice. I could see Jenny's mom crying in the front row as Chris took Jane's hand in marriage.

We retired to the tent after the wedding for the reception. Chris and I had run power cords from the barn to the tent so that we could have a DJ and lights. Chris had hired a local guy to roast a pig, which had started at about 5am that morning. You could smell the roast pig throughout the farm. Jessica and Jane, with the help of her parents, had been cooking for days to prepare the sides. They had a friend make their cake since there wasn't a bakery in town.

They had an open bar, which two of Chris friends had volunteered to man. They were mixing drinks and handing out beer and wine to the guests as fast as they could. Overall, the wedding and reception ended up costing about $5,000 since more of it was done in house. It was surprisingly inexpensive compared to most wedding today, but it was a beautiful and meaningful event since it was all homemade, hand-picked, and much of it was from the farm itself.

While I was waiting in line to get another drink at the bar, I ran into Don who is Jenny's uncle. He is a union leader in town and had spent most of his life supporting the union. He was a big guy with large hands that could build just about anything.

"This is a nice event," I said making small talk.

"It really is. Great, to have a downhome wedding. It seems like costs on everything are going up lately and it is hard to afford most things today the way that we could when I was younger," Don said.

"It definitely seems like it is harder to make ends meet these days,"

"I read an article the other day that said that since I was in college in 1978, college tuition has gone up 318%, medical costs have increased by 378% while average pay has decreased by 32% [1]. It's a real shame and it's not that surprising that so many of us have a hard time making ends meet."

"I know. A lot of kids are going into debt from college, health care is expensive, and the cost for food and a mortgage just about empties out our banking account. It definitely feels like times are hard," I said.

"That is because most of the money goes to the wealthy, like the CEO of a company, and the workers earn very little money. It's not a good way to business or to maintain a stable society," Don said.

"I agree. It seems that we need to raise the minimum wage, require caps on CEO salaries, and tax the wealthy a fair share for the benefits they receive from the system. That really seems to have gotten out of hand in the US."

"We keep hearing that if small businesses and CEOs make more money, then they will pass those earnings on to the workers and everyone will improve, but I have never seen any extra income passed on to the workers. It seems like the rich just accumulate more money, and they don't even spend it. Where if someone in the middle or lower classes get more money, they usually spend it quickly because they are in need of basic goods and services. That's the way to grow an economy, but no one in the US, at least no one in power, seems to get that or at least is not willing to fight to make it a reality," Don said.

"I would like to see those changes as well. I am surprised that more people in power don't support a $15 minimum wage. That would definitely increase the spending power of the lower and middle classes and as you mentioned, that money is likely to be turned back into the local community pretty quickly, raising everyone's ship in the middle class economy," I said.

"Yeah, it's a good plan, but we can't seem to get the rich to support it," Don said.

I finally made it up the bar and got another drink for Jenny and myself. "Well, it has been good chatting with you Don. Hopefully, we will get to the place where all Americans can make a better wage and have more disposable income," I said as I departed and went back to my seat next to Jenny at the head table.

Chapter 16

The Hoosier Environmental Council

Indianapolis, Indiana
October 5

I was having breakfast and coffee with Jenny at the kitchen table when she said "With your article on sustainability, you should come into the office with me and interview all of my coworkers to see what they're doing. Our office deals with the state policies and lawmakers for any bill that affects the environment or sustainability. It would be a good background piece for your article."

"That's a good idea. I actually don't have any interviews planned for today. Would it work to come in today?" I asked.

"That would be great. I think that everyone will be in the office, so you'll be able to talk with all of our divisions and see what we're doing," she said.

I drove the Prius from our house in the outskirts of Indianapolis to the downtown office of the Hoosier Environmental Council [2].

We parked in a parking garage in the downtown area and walked a few blocks to the office near the state house. The Hoosier Environmental Council spent most of its time working with legislators trying to influence bills and being an environmental watchdog. They had to be ready to run over to state house on a moment's notice if an important bill came up for a hearing.

"Hello Candace," Jenny said to the front desk secretary as we walked in. "This is my husband, Greg, who will be visiting with us today. Can I get a guest pass for him?"

"Sure. Just a minute. I assume your last name is the same as Jenny's. That's Cunningham right?" she asked.

"Yes, that's correct," I said.

"OK, let me get this into the computer and print out the badge. You'll need to wear this at all times, but it also gets you access into the court house if you need to go over there for business," she said.

"Great thanks!" I said. It took a minute to get the name badge, but then I was all set.

The offices were tight and the group was squeezed into the space. But it was in a nice limestone building. "Now that I think of it, I'm surprised that I've never come to work with you before," I said with a smile. It was nice to see where Jenny worked and what she did.

We headed towards Jenny's desk and passed a few cubicles.

"This is Sebastian. He covers the climate change desk. Sebastian, this is my husband Greg," she said.

"Nice to meet you. I liked your article in the New York Times," he said recognizing my name.

"Thanks."

"This is Amanda Reynolds; she covers clean energy for us. Amanda, this is my husband Greg," Jenny said.

"Nice to meet you. Jenny, tells me that you are working on a current bill before the Indiana state legislature called SB-309," I said.

"Yes, it passed into law last summer, but we're now working on its repeal," Amanda said. "It actually will not allow future net metering of solar installation into investment-owned utility companies."

"What does that mean?" I asked.

"If you don't currently have solar installed, you can't get paid by producing extra energy that flow back to the grid. This is a very backward direction for our legislature to go. In 2015, US solar energy jobs overtook oil and natural gas extraction jobs for the first time in history [3]. Coal jobs and production have been declining in the United States since 2008. We currently only employ about 86,000 in coal where it had peaked at 883,000 employed in coal in 1923 [4]. Wind employs 102,000 people and solar is the strongest in the electrical industry employing 374,000 people in 2016 [1]."

"I'm surprised that solar and even wind are so strong in the electrical industry. During the Trump and Clinton debates, all that we heard about was bringing coal back. But many people pointed out that coal had been in a long-term decline because natural gas was so cheap. I never knew that wind and solar were so much stronger in the US economy than oil and natural gas," I said.

"Solar is a quickly growing energy sector and the Indiana State legislature just voted to take the state out of the solar game. Without net metering, there's not as much value for homeowners to install panels. They would have to get battery backup or just use the energy as it's produced. Most houses produce energy during the day, which flows back to the grid and is used by industry. Then at night, other electrons flow into the home providing light in the evening," she said.

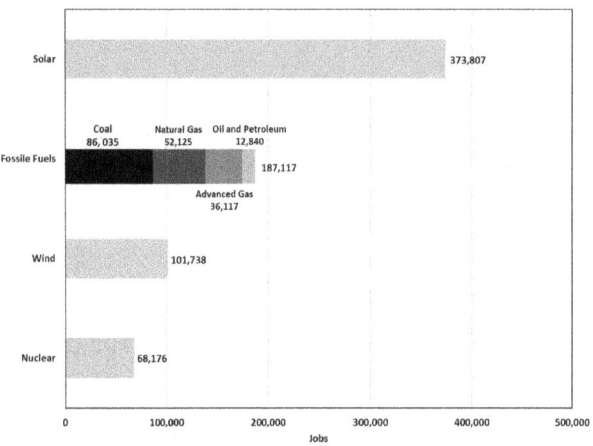

Solar employs more workers than all of the fossil fuel industry for energy generation. Data from US Department of Energy. [1]

"It sounds like this law would kill the future development of solar power in Indiana, which can't be good for Hoosiers," I said.

"It would kill the home solar industry and be very bad for Hoosiers," she said.

"So why did the legislature vote this into law?" I asked.

"The fossil fuel lobby has a strong voice in the state house in Indiana and made sure that future clean energy efforts are restricted," she said.

"But that doesn't make any sense. It's clear that solar is growing from the job statistics. It seems like a tide that can't be stopped," I said.

"Yet the legislature is at least trying to slow that tide. Any home solar that is installed after 2022 would not be able to tie into the grid which would kill the future of solar in Indiana."

"Well, good luck with the repeal. Hopefully that will happen before the law comes into practice," I said.

"We were shocked that the state senators would even propose this bill not to mention actually support it," Jenny said. "The only people supporting the bill are the coal industry but they have a lot of influence in the state."

"I guess so. It's too bad a special interest can shape the future of alternative industry for the entire state," I said. We took our leave of Amanda and moved further into the office suite.

"This is Rachel; she covers the natural resources desk," Jenny said.

"Hello Greg, Jenny talks a lot about you," Rachel said.

"Thanks. What are you working on in natural resources?" I asked.

"We're helping to support SB-420, which if it passes would protect older forests in Indiana and require that the Department of Natural Resources, called the DNR for short, put aside 10% of their larger forest holding for long-term protection. DNR biologists suggested this very thing in 2005 and current biologists and senators support this bill, but the senate chairman won't bring it to a vote and DNR opposes it now," she said.

"Why is that?" I asked.

"In 2005, the DNR changed its approach to its managed forests. They started to log these forests heavily which is what prompted DNR biologists to suggest putting 10% of the older forests aside for preservation. Now they make a good bit of money from the timber sales and it seems that they don't want to be restricted by a law like this," she said.

"And this is Julie; she covers water resources. This is Greg," Jenny said.

"Hey Greg," Julie said.

"Hello Julie. What is happening with water in Indiana?" I asked.

"Well, coal ash ponds are located right next to waterways that flood on a regular basis. Very little is done to make sure that they don't contaminate our waterways. Excess agriculture nutrients are filling our streams and causing algal blooms locally and all the way down to the dead zone in the Gulf of Mexico when it's added to the rest of the waters flowing down the Mississippi. And most of our waterways are contaminated with *E. coli*, which is an infectious disease, probably from animal manure," she said, listing off a series of local issues they were dealing with.

"That's a lot to keep up with," I said.

"There's constantly something that is coming up that threatens the environment in Indiana. The state is not watching these issues so non-governmental-organizations like ourselves have to be there to try to keep industry in check so that it doesn't destroy our waters, natural resources, and air for future generations. It's a lot of work to undertake," Julie said.

"Thanks for the overview," I said in parting.

"Well, that is most of our staff at the Hoosier Environmental Council," Jenny said.

"I should have been aware before, but I'm constantly surprised by everything that goes into sustainability. Just walking through your office, you hear about a whole suite of issues that most people would not think are related, but they all come together with protection of the environment," I said.

"That's true and we haven't even gotten into the environmental justice issues. We all deal with environmental justice issues and keep an eye out for when minority populations bear a greater pollution burden than the average resident. It's surprising how often that's the case. There is certainly a lot of work to do, just to keep industry and the state from trampling all over the citizens' rights to clean land, water, and air. And these are pretty basic rights. Somehow, the industries think that they have a greater right to these resources," Jenny said.

"Those are externalities to the calculation of the goods and services. If industry actually had to pick up the bill for the pollution that they caused or the human health issues that resulted from their pollution, it would be much clearer where the responsibility lies for many of these issues. Individual products would be more expensive, but we would be paying the true cost of those products rather than hiding the cost for some products in the destruction of the environment or degradation of human health," I said.

Chapter 17

AASHE Conference

October 15
Portland, Oregon

The Association for the Advancement of Sustainability in Higher Education [1] conference was being held in Portland, so I decided to see what this national organization for sustainability professionals had to say on the topic of sustainability. Obviously, they knew a lot about it.

Portland itself is a lesson in sustainability. Their public transportation system is superb. You can travel anywhere in the city on their light rail system called MAX for Metropolitan Area Express [2]. The city has developed an ordinance that does not allow the city to sprawl anymore, which means that it can only infill or go up with higher buildings. This keeps the city compact and makes it navigable. It is interesting to see a city limit itself this way.

I had arranged to meet a thought leader in the field of sustainability named Dr. Ernesto Campione to get an insight into sustainability and what we know of the concept. He had been working in the field for 20 years and helped to develop our basic understanding about sustainability. We sat down to talk in the main lobby of the conference hotel.

"Dr. Campione, how are you?" I asked.

"Greg, good to see you. I'm doing well; and you?"

"I have just been enjoying Portland. It's fun to travel in a city with such good public transportation. You can get around with ease and get a glimpse of what the future will be like for all of us," I said.

"Yes, large cities and public transportation seem to be the key to many solutions in sustainability. It'll be important for much of our society to live in more dense human population centers so that we can concentrate resources and not have to drive everywhere to acquire them," he said.

"I'd heard that urban concentrations of population were one idea for making the human systems more sustainable," I said.

"In the next 30 years, we expect to add 2 billion people to the global human population and most of them are likely to live in urban areas. This minimizes our impact on the landscape, we can concentrate resources for people to use, and surrounding countryside can be used for intensive food production. Resources that we are thinking about are things like grocery stores, parks, fitness centers, and schools. We don't have to build as many of them or bus people around town if we have high-rise apartments for people to live in. Then all of these people would support a local economy of shops that are walking distance from the apartments," he said.

"If food is produced in the surrounding countryside, then that would reduce food miles and everyone would be eating fresher food that's local," I said.

"Yes, it becomes an efficient way to live well on the planet with less environmental destruction. We went through a period of urban sprawl in the United States where many people moved to the suburbs. Then people became dependent on their cars and could not walk to most places that they needed to go, like the grocery store or your children's school. Now, we see the opposite effect. We are contracting back into cities and concentrating our population," he said.

"What have you seen at the conference that you're excited about?" I asked.

"I just came from a session on the Post Landfill Action Network [3], called PLAN for short. They focus on reusing items on college campuses and capturing all of the materials that are thrown away at move out. Student organizations can collect those materials and store them over the summer. Then the students can sell those items back to students at cheap prices, but enough to support the organization and the student workers. This provides good cheap items for the students, removes waste from the landfill, and provides sustainability jobs to students on campus. It's a very effective sustainability program that's now on many college campuses," he said.

"Yes, I saw that presentation as well. It seems that the PLAN members are also providing student organizational training now to get students to self-organize to make things like this happen. It's a good approach that naturally diverts useful items from the landfill," I said.

"I saw a presentation by Willamette University on its Green Rotating Fund [4]. They charge a small tuition tax that the students put in place to collect about twenty dollars from each student on campus. This money creates a fund that is approximately $100,000 a year. They use this fund to provide micro grants to students that have ideas for sustainability solutions on campus. Then any revenue or profit from those solutions comes back to the fund to provide opportunities for future students. Through this whole process, sustainability initiatives are being completed and the campus is becoming more sustainable. Also, the students manage the fund and students have to submit grant applications to get money to do their own projects. So many students get experience managing large amounts of money, writing grants, and delivering on grants. It seems like the most efficient way to fund sustainability initiatives on a college campus and also provides great learning opportunities," he said.

"That does sound like a great program. I like how the best sustainability solutions solve a problem, but also provide greater benefits like experiential learning and life skills," I said.

"The best talk that I've seen for big picture thinking in sustainability was the Natural Step program [5]. This is another way to frame sustainability. A Swedish scientist, Karl-Henrik Robèrt, had the idea in 1989 to bring together scientists, engineers, and mathematicians to discuss earth as a system and determine what we can from a scientific perspective of the earth. They started with basic principles like the Earth being an open system in relation to energy and a closed system in relation to matter. Thinking through some of the basic laws of physics, such as that matter can't be created or destroyed, only transformed and that all things tend to disorder, the group was able to make some basic recommendations about the Earth. These four recommendations were to reduce our dependence on fossil fuels and heavy metals, reduce our dependence on synthetic chemicals that persist in nature, reduce our destruction of nature, and ensure that we are not stopping people around the world from meeting their basic needs. These concepts are the building blocks to a sustainable future," he said.

"Some of this relates back to my last article called Exposé on Climate Change. It showed how we were excessively extracting fossil fuels from the lithosphere with the pure motivation of profit rather than thinking about the long-term consequences. Also the second principle of natural step, to reduce synthetic chemicals, is a clear need as well. We are producing plastics at a great rate and nature is not equipped to break these produces down. We also produce other chemicals like artificial herbicides, pesticides, and other chemicals that aren't found in nature. All of these play a role in diminishing the environment and ultimately human health," I said.

"Right, these concepts also tie in with the United Nations Millennium Development Goals [6] that were established in 2000 and 191 countries signed on to them. These goals were meant to be achieved by 2015 and were the following: Eradicate extreme poverty and hunger, achieve universal primary education, promote gender equality and empower women, reduce child mortality, improve maternal health, combat HIV/Aids and malaria, ensure environmental sustainability, and develop a global partnership for development. These were truly lofty goals to bring the entire world up to better living conditions for all humans. Of course, 2015 has come and gone and progress towards these goals has varied by country but progress was made in most areas. In 2016, these goals were changed to the Sustainable Development Goals which expanded to 17 global goals that embodied many of the same principles, but added some others like taking strong action to combat climate change [7]," he said.

"The United Nations has deeply embraced sustainability as a core principle," I said.

"They truly have. All of the countries around the world come together on a regular basis to discuss these issues. A lot of work is being conducted to move towards these goals. Of course, they are hard things to achieve but definitely well worth the effort," he said.

"Well thank you for your time. This has helped me to see the bigger picture related to sustainability initiatives and theory," I said.

"You're welcome. Have a good conference," he said in parting.

Chapter 18

Confrontation at Home

Robertsville, Indiana
October 22

Jenny and I were enjoying a quiet Saturday morning at home sitting on our front porch. The fall leaves had changed colors and the woods were painted with hues of brown, gold, and red. The temperature was warmer than it should be for October, but the number of insects had started to decrease. Jenny was eight months pregnant and we were looking forward to our baby's birth soon. We were going to have a boy and had arranged with a doula and a midwife to give birth at the local birthing center.

We were still discussing boy's names when someone drove up to our house. He looked familiar, but I was having a hard time placing him.

"Hey Greg, I hope I'm not bothering you," he called up when he got out of the car.

It took me a couple of seconds to place him, but after a pause I said, "You're the person I met at ISU at the walkability tour, right?"

"Yes, James Bedford. We met during the tour. I was interested in your article and was wondering if I could be of any help, so I thought I'd stop by," he said.

"What did you say that you did again?" I asked. I still did not trust him and having him figure out where I lived was disconcerting.

"I'm the sustainability coordinator for the city of Jasper," he said.

"But Jasper is more than two hours away and I wouldn't think that a town of 15,000 people would hire a sustainability director," I said.

"I had some business in Indianapolis, so I was close by. Do you mind if I come up? This must be your wife, Jenny?" he said.

"Yes, Jenny this is James Bedford the sustainability coordinator for Jasper, Indiana. Do you mind excusing us while we talk about sustainability?" I asked.

"No, I'm happy to go do some work if y'all want to talk shop," Jenny said and went into the house.

"Nice place that you have here," James said.

"Thanks. What did you want to talk about?" I replied.

"I was just wondering where you were in your story and if I could provide any assistance since I know sustainability well," he said.

"As I mentioned at ISU, I'm investigating sustainability initiatives for an article for the New York Times. I'm trying to get an overview of what sustainability means to its practitioners. Sustainability is a term that we use a lot, but I find that most people don't really know what it means," I said.

"Yes, I find that's true in my experience as well," he said.

"So how would you define sustainability? I asked.

"Well, you know, recycling. That's mainly what I do is run recycling programs," he said.

"But isn't sustainability so much more than recycling? What else do you do?" I asked.

"Well, I have meetings and talk to people. But it's all about recycling," he said.

That was the worst summary of sustainability that I had ever heard from someone who claimed to work in the field of sustainability, so now I did not trust him at all.

"How did you find out where I live or even who I was for that matter? I never introduced myself. You also seemed to know who my wife was without introduction," I said.

"I asked the conference coordinators about you and then did some research online. That's all. I just wanted to connect with you and talk about your article," he said in a placating manner.

"What do you want to know about the article?" I asked.

"I thought it sounded interesting so I was wondering what you were including in it. So what are you putting in your article?"

"That really isn't any of your business. I think that it's time for you to leave," I said.

"Listen Greg, you're messing with things that are bigger than you. You had best let this article go or you might not like the consequences."

"Are you threatening me?"

"Not at all. It's just that we know what you did to the Extreme Oil Company and my employers aren't going to stand aside and be the next victims of your vicious articles," he said.

"I have the right to investigate this story and free speech gives me the right to report on it to the public."

"You might have the right, but I don't think that you want to go there."

"What do you mean?"

"With such a nice wife and a child on the way, you don't want to leave them without a father," he said.

That stopped me cold. I stared into his eyes and he didn't blink or look away. I could tell that I was looking into the eyes of a killer.

"I think you've said enough. It's time for me to call the police and report this conversation," I said as I took out my phone and dialed 911 with my finger over the call button.

"Just think about it. We don't want it to end that way, but we think it is time that you end this story or we will have to," he said as he walked down the steps and headed to his car.

I watched him back out of the driveway and memorized his license plate number as he pulled out onto the street.

I went inside once he drove away.

"Did you hear that conversation? He just threatened me if I don't stop writing this article. He's the same guy that was at the sustainability conference at ISU and he tracked me down," I said.

"I remember you talking about him and thought it was odd when you sent me inside, so I videotaped the conversation through the window," she said.

"That was good thinking. Can you send the video to me? I wonder if you picked up what was said through the open window?" I said. "I should probably get a restraining order against him."

Chapter 19

Jones, Hackett, and Williamson

Robertsville, Indiana
End of October

I set up a Skype call with John Hackett who was with the law firm Jones, Hackett, and Williamson. He was the lawyer that I worked with against the oil companies and they received part of a very nice settlement. I figured they would be willing to give me some pro-bono advice and maybe set up for another possible lawsuit.

"Hey John, can you hear me?" I asked.

"Yep, and your video is coming in just fine. How are you doing?"

"I'm well."

"I hear you have a baby on the way. Congratulations!"

"Thanks! It's due in a couple of weeks. We're really excited and nervous at the same time," I said.

"It's nerve wracking, especially the first one. We have two kids ourselves that are now six and eight years old, but that first birth is scary. You don't really know what to expect and the mom isn't sure how she will deal with the pain. It is a big challenge. I'm so proud of my wife and I saw this inner strength that I wasn't really aware of prior to the birth. It's amazing how strong women are and how much pain they can endure. I wouldn't want to be the one giving birth. What did you want to talk about?" John said.

"I have a new exposé that I'm working on. This one is looking into sustainability. It was motivated by a MallMart moving into our small town and running all of the local markets out of business with their low prices and competitive tactics. The article as a whole is looking at all aspects of sustainability so that the reader will get a better understanding of the breadth of sustainability by the time they are done reading, but MallMart seems fixated on what is in the story. They sent a guy over to my house who threatened me to end the story. I just wanted to get some back up from you and also get some insight into what a business can be doing that would make them so panicky," I said.

"As usual, you are poking the hornets' nest. I guess that's how you get to the bottom of important stories. Off the top of my head, a company can be doing many illegal things to get a competitive advantage. A company like MallMart is at the top of the game and it always seems that these big businesses and powerful people have the greatest distance to fall. They could be doing insider trading by giving away corporate secrets about their actions for others to benefit from stock trades based on that information. They could be price fixing in collaboration with their providers to make unfair deals to benefit one company. They could be breaking all sorts of environmental laws to get cheaper prices for their products. The list is really quite long. And then, they can do some of the things that normally land you in jail like attacking people or trying to kill them," John said.

"That last bit is what I'm most concerned about. I thought this was going to be a simple story about sustainability and how people make daily choices and companies can contribute to or detract from sustainability. But the level of interest that MallMart is paying to my story makes me think that they're guilty of some bigger things, so I'm pushing deeper on their business practices now after the threats," I said.

"It's interesting how their guilt and pressure might result in their demise more than their original practices." "Yeah, it's ironic," I said.

"I've got a friend in the Indianapolis police force. He was an FBI agent that I worked with and decided to move back home to be a detective in Indy. Why don't I give him a call and have him connect with you. You can tell him your story and concerns. Then you can call him if anything comes up."

"That would be great. It'll give me some piece of mind to know that I have backup out here." "Now what about this case if it goes forward. Is there something that you would need our services for?"

"I hope so," I said as I explained what I had found so far and laid out my evidence.

We decided to stay in touch as the story progressed and John committed some of his firm's time into looking into the public record on MallMart.

"I'll share what our paralegals find with you for your story. If we're lucky we can build a case and we can provide good information for your story at the same time. Give me a call if anything happens on your end and my friend from the Indy police force will be in touch in the next few days. Good luck with the birth and let me know how that goes as well."

"I will," I said as I clicked the red phone icon on my screen.

Chapter 20

The Birth

Roberstville, Indiana
November 12-13

The baby was due on November 2nd and is often the case, it wasn't very good at marking the calendar. That date came and went and still the baby was not ready to be born.

"I actually enjoyed being pregnant after the first trimester of morning sickness. But I am ready for the baby to be here. It's hard just getting up and down from a chair and there's no way that I could pick something up off the floor," Jenny said.

"I'm ready to have the baby born, too. I have to say that I'm a bit nervous about the birth. It seems like such a difficult process. We've read all the books and watched the videos. We prepared his room and have the right car seat to bring him home from the birthing center. I don't think that there is anything else we can do to prepare," I said.

"Our doula is on call and we are set to go to the birthing center where the midwife is on call. We have the inflatable pool for water labor. All of the tests looked good and the baby is in a good position. So we just have to wait until my body is ready to give birth to the child," she said.

"It would probably help if we decide on his name before we go to the hospital," I said.

"Yeah, I know. It's just so hard to decide. David is a family name, but I always liked Joseph, and Ezra would be more distinctive," she said.

"It does seem that people are preferring names that are a bit unusual. I remember when I was in fifth grade there were four girls named Jen. Of course, you can't know for sure what our child's future schoolmates will be named, but I want to avoid the top ten list of names right now," I said.

"It's a big responsibility to determine what someone else is going to be called for the rest of their life."

It was 11:13pm, of course, on November 12th when Jenny's contractions started. I was already asleep for the night.

Jenny nudged me and said, "I think it's time. I'm having contractions but my water hasn't broken yet."

I was groggy for a second, and then what she had said sank in. I jumped out of bed and helped her get up out of bed and get dressed. We had packed to-go bags so that we were ready and everything else, like the inflatable pool and the child seat, were already in the car.

We grabbed our to-go bags and made our way to the car. It was a bit surreal as we tried to rush to the car but could not walk faster than a slow gait because of Jenny's advanced stage of pregnancy.

I called our doula, who answered right away and said that she would meet us at the birthing center. She said that she would also call our midwife to alert her we would be arriving soon.

We made it to the car and got Jenny in the passenger seat. I threw our to-go bags in the back. The drive to the birthing center was just 10 minutes from our house on the north side of Indianapolis.

"How're the contractions?" I asked as I drove fast, but carefully.

"Still pretty far apart. I think we have some time," she said. I could see her wincing through contractions and breathing like we learned in Lamaze classes.

I parked at the birthing center, grabbed our bags, and helped Jenny from the car. We made our slow progress to the front doors and our doula pulled up just then. She hurried over and helped Jenny get through the front doors.

"How're the contractions?" she asked.

"They're strong but pretty far apart. My water hasn't broken yet," Jenny said.

"I called the midwife and she should be here soon. The staff can get you checked in and get you a room," the doula said.

We checked in and the nurses hurried Jenny back to a private room. After completing insurance paperwork, I was allowed to go back and see her. I dropped our bags in the room and checked on Jenny. I set up my small speaker and music on my iPhone. We had created a classical music play list that was supposed to help Jenny relax. It was playing quietly in the background and helped me relax a bit.

"She's doing fine. Do you want to set up the tub for laboring?" the doula asked.

"Oh right," I said and rushed out to get the stuff from the car.

We had seen online that many people suggested laboring in water, but most facilities did not have a pool for this purpose. So we had to get our own inflatable small pool that was about six feet in diameter and a water pump. We brought along a hose that we could connect to the faucet for filling and the pump was for removing the water down the drain after we were done. It turns out there is an underground network of people that get this equipment and then it's passed from mother to mother. The state of Indiana allows laboring in water, but not giving birth in water. The water is supposed to help support the woman and her baby during labor, which is better than laying on the bed. It's hard to get comfortable between contractions and the water helps with that.

It was a bit comical if you were watching this process from the outside. I was flustered from the impending birth and trying to get all of this pool equipment picked up and carried into our room in the birthing center. Then, like Tim Allen on the TV show, I worked at setting up the plumbing for the tub and then filling the tub with warm water. Honestly, the manual labor of using my hands to do this work and be useful helped to take my mind off my wife's obvious pain.

We wanted to have a natural birth and had written a birthing plan, which we gave to the doula and the midwife. Jenny opted not to have any pain medication because she felt that it would interfere with her body's signals for what to do during the birth and we wanted to avoid a Cesarean Section.

The midwife came in when I was half way through filling the tub. She went right over and checked on Jenny. She measured her dilation.

"Jenny, you're at 3cm dilation, so you still have a ways to go. It might help if you walk the halls for a little while," the midwife said.

I had finished filling the tub, so I came over to help Jenny get out of bed. We slowly walked the halls with the doula, who was reminding Jenny about breathing exercises and trying to relax.

About a half hour into our walking circuits, Jenny's water broke. We went back to our room and changed Jenny's gown while some the nursing staff cleaned up the hallway.

The midwife came in and said that Jenny was progressing well. "Your contractions are coming closer together and you're continuing to dilate."

"Can I get into the pool?" Jenny asked.

"Let me check the water temperature," the nurse said pulling out a small thermometer from a breast pocket. "It's a bit cool. Greg, do you mind exchanging some of the water and adding some hot water."

I quickly did this, bringing the water temperature back up close to body temperature. Then we worked to get Jenny in the tub.

She squatted down in the water and sort of leaned back against the wall of the tub.

"This's much better," Jenny said between contractions.

She spent the next hour in the tub moving to different positions. The doula put pressure on her lower back and massaged her back on occasion, which seemed to help Jenny relax between contractions.

I could tell that the contractions were painful, but Jenny was dealing well with the pain. She seemed to be in her own meditative state, just staring off in the distance when she wasn't clenched up in a contraction.

The midwife checked her a few times in the tub. I replaced some of the water a couple of times, but mostly stayed by Jenny's side, holding her hand and encouraging her. I was so proud of her strength and bravery.

Jenny's labor continued for two more hours in the tub. The contractions were coming close together now and the midwife declared that Jenny was 10cm dilated. She said that Jenny had to get out of the tub and move to the bed at this point. We helped Jenny from the tub in between contractions and quickly swapped her gown out for a dry one.

Jenny made it into the bed just before the next contraction.

She was concentrating so hard at this point that she did not seem to register anything else going on in the room.

The doula stayed by Jenny's side and was saying encouraging words to her through the process and helping her relax in between contractions. The midwife was with us all the time now.

Jenny's contractions got really strong and the midwife started to encourage Jenny to push through the contractions. I was holding Jenny's hand and it felt like she was going to crush my hand during the contractions. With about five strong pushes through the contractions, the midwife finally said, "It's a baby boy. Congratulations!"

I was amazed to look down on my baby boy all wet and wrinkled with dark black hair. He started to cry at the top of its lungs.

The midwife cut the umbilical cord, wiped off the baby a little bit, and gave him to Jenny to hold. He stopped crying right away and snuggled into her arms. Jenny looked worn out but happy to hold her newborn son.

"What is his name?" the midwife asked.

Jenny looked at me and said, "David Joseph Cunningham," and the nurses wrote it down on the official forms.

Jenny expelled the afterbirth a little while later and the nurses came and cleaned up the baby and measured his vital signs and weighed him in the room. I got to hold him while they changed the bedding and Jenny's gown one final time.

"David looks very healthy. All of his vital signs are good. He's 7 lbs and 5 ounces. Congratulations again," she said as she left the room.

I was so happy to be a father and so proud of Jenny's strength going through this most human experience. I couldn't help but think that this is what it truly means to be sustainable. The bigger biological goal is to pass your genes on to the next generation, but we also have to think about passing on a world that's not in worse condition than the one that we inherited from our parents.

Chapter 21

Cradle to Cradle

Mountain Lake Biological Station, Virginia
December 11

About a month after David's birth I went to Charlotte, Virginia to meet with William MacIntire. He was a US architect that had changed the way that we think about industry, manufacturing, and design.

I wasn't ready to leave my family for a week, but I had to continue my interviews and this conference was just a day's drive away. It was hard to leave Jenny and David for this time, but Jenny assured me that she would be OK. Her parents were coming over to stay while I was gone to help her with the baby. At this point neither Jenny nor I were getting much sleep. I hoped to catch up on a little bit of that sleep while I was travelling so that I could be coherent during my interviews.

William MacIntire had proposed the idea that we need to stop thinking about how we make things as a through-put society in which products end up in a landfill at the end of their use. We actually need to manufacture things in a way where the end product is the feedstock for the next generation of products. This was a radical idea, because it would completely change the way that we do manufacturing. It was so far ahead of its time that it took most of a decade to see companies actually using the concepts to their advantage.

I met William at the Mountain Lake Biological Station, which was part of University of Virginia. The field station was behind the resort where the movie *Dirty Dancing* was filmed in 1987. It was set up as a miniature of the University of Virginia campus with cabins around a central grassy square and the main teaching building and laboratories at the head of the grassy quad. The entire field station was set in the Appalachian Mountains amongst the eastern hardwood forest. On the ridge tops there were a few table mountain pines that are scruffy-looking stunted pine trees that could be hundreds of years old.

William was at a conference at the field station on sustainable design that included architects and interior designers. He had just finished a talk on his signature book, *Cradle to Cradle* from 2002 [1] and his more recent book, *The Upcycle: Beyond Sustainability-Designing for Abundance* [2] when I caught up with him.

"Mr. MacIntire, nice to meet you," I said.

"Thanks. Good to meet you Greg. I've always appreciated the reporting of New York Times; I'm happy to talk to you about sustainability," he said.

"I enjoyed your talk. It seems like your ideas are finally catching on in the industry and companies are starting to use your ideas for sustainability," I said.

"It took a while, but then it does take a while to spin up and fund a company from ground zero. I'm happy to see some companies using our concepts and it seems to be working for them. They sell products, once they wear out and need to be replaced, the company takes them back and breaks them down so that they can make the next generation of products. It's a relatively simple concept that mimics the processes of nature," he said.

"What do you see in nature that you are trying to copy?" I asked.

"There's no waste in nature, so there should be no waste in industry or in our human lives. The waste that we do produce is just an economic burden on the company and an environmental burden on the planet. Therefore, it just doesn't make business sense to continue producing the way that we have been. It wastes resources and wastes money. There are better ways to do manufacturing and that is what we show in our writing," he said.

"What are some examples of companies using this concept of no waste?" I asked.

"Well, check out this carpet here," he said, pointing to the carpet that looked relatively new outside the main presentation hall. "These are carpet squares about 2 foot by 2 foot. They are manufactured by a company called Interface that has been working since 1994 to reduce its burden on the environment. They have improved their product over the years so that now it's the gold standard in sustainable flooring. They don't put off the environmental burden to the next generation. They say that they're more in the carpet lease business than the carpet sales business. They sell you these carpet tiles and one of their certified companies comes out to install it. Then if you get a stain on a section of carpet that can't be cleaned or if a section of carpet wears out from frequent use, the installation company comes out, removes that one or a few carpet tiles, and replaces them with the same type. They take the waste carpet and peel off the top stain resistant fabric, peel off the under-padding, and then peel off the support frame. All of those pieces get shipped back to the company and melted back down as feedstock for the next generation of carpet tiles. You get a brand-new looking carpet and the company gets more material that fits their specifications for their manufacturing process."

"That really is amazing. There's no waste and the product can be made again with a little energy input. The company also keeps a customer with the smaller cost of replacing just the damaged tiles so the customer also saves money," I said.

"It truly is a win-win situation and all that it took is some prior planning and engineering on Interface's side to make products that could be recycled like that," he said.

"That's a big step towards sustainability, but good environmental practices are not everything," I said.

"No that's true, and the company embraces all of the aspects of sustainability. Many of these companies will hire employees, train them, have them work for a while, and then fire them when manufacturing gets slow or it's the off-season for the product. But Interface makes the commitment to maintain good employees, invest in their training, and keep them on the payroll even in hard times. They will actually pay for their employees to spend their time volunteering in the community rather than work during slow times. That keeps them on the payroll and also provides the perception of social benefit to the company from the community that they work in. Also, the company gets the benefit of keeping well-trained employees that are now loyal to the company because the company has their backs," he said.

"What about the financial side?" I asked.

"They started with a good business plan that had enough funds in it to get through the beginning slow times as their company was getting recognized for the good work that they do. As the company got more established, they slowly hired more employees, making sure not to overextend. Now they are a mature and thriving company. The company makes a profit, which allows it to stay in business; the employees have secure jobs and they also get to give back to their communities during slow economic times. And all of this helps the environment with no waste," he said.

"So they truly are a real sustainable model for a business. Are there any others?" I asked.

"Another company is called DIRTT which stands for Do It Right This Time. They create modular wall treatments, from natural wood wall covering to green walls with living plants. They use Forest Stewardship Council certified wood products. That means that their wood products are sustainably harvested with a mind towards sustainability of the forest system from which the products come. Often times, they are planting trees as well to maintain the forests where they harvest," he said.

"That's cool. I didn't know that there was an organization that monitored how wood was harvested and only supported sustainable harvesting with their seal of approval. What other things does DIRTT do?" I asked.

"They also conduct a Life Cycle Analysis to determine the environmental cost of harvesting, using, and then ending the life-cycle of that product. All of this weighs into the cost of the product and the cost to the environment so it's less costly if the original product can be reused or recycled rather than discarded in a landfill," he said.

"That's great that they monitor the full costs of their products," I said.

"Using their products in your new construction or renovation projects get you points in LEED certification and they are also an important part of the Living Building Challenge," he said.

"What is the Living Building Challenge?" I asked.

"It's a building standard that is beyond LEED certification. LEED follows a checklist of points that you can get for a series of sustainable solutions. Some people game that system by doing the minimum to get enough points to call a building LEED certified, but they don't think innovatively to push to new building processes. The Living Building Challenge has 20 imperatives that you must follow. These include providing more energy that you consume, providing clean water for the building and the rest of the natural system, providing space for urban agriculture, and other criteria such as Beauty, Inspiration, and Education. The Living Building Challenge is pushing the industry to do better and make structures that give back to the environment rather than simply taking from the environment. They use the metaphor of a flower that provides beauty, but also resources for the rest of the system. That is what they envision for their buildings," William said.

"Thanks for the information. I'm realizing that there's a lot more to sustainability than even I thought at the beginning. There are a lot of innovators that are active in coming up with solutions that are much more environmentally friendly," I said.

"Yes, their ingenuity is impressive. Through these initiatives, they are findings the solutions for the next generation of best practices for construction. They have already started to change the way that we build and decorate buildings. I think that it will only continue to improve and the end results will be buildings that improve our health rather than takes away from it," he said in parting.

Chapter 22

Rephrasing the Conversation

Mountain Lake Biological Station
December 12

The next day, I met with another researcher named Per Esper Stokes who was a psychologist and economist from Lund University in Sweden [1]. He was one of the leading thought leaders on how we rephrase how we discuss climate change so that we don't create fear, denial, and distance from the problems that we face.

"Hello Dr. Stokes. It's nice to meet you," I said as I shook hands with the tall and lanky dark-haired Swede.

"Hello Greg. It's nice to meet you. Please call me Per. Come, have a seat," he said as he led me over to a couple of rocking chairs next to a roaring fire. We were in a sitting room with a view out to the hardwood forest outside. It was well into December, but there had not been any snow yet this year at the field station. The temperatures were warmer than usual and we only needed a light jacket to walk around outside.

"I understand that you've been working to study the psychology of human responses to climate change along with the economic decisions that people make," I said.

"Yes. It's an interesting phenomenon. In the 1990s, scientists were talking about climate change and had studied it enough that they knew that it was happening. Back then it was vogue to put up the front of being purely objective. So scientists would work in their labs, publish their papers, and maybe talk at scientific conferences to other scientists. But this approach was not getting the word out about what they were seeing in their data. And the more time that passed, the more concerned scientists became about climate change. So in the 2000s, scientists realized that they had an obligation to inform the public because much of their research was funded by public funds through tax dollars to the national scientific organizations like your National Science Foundation. They decided that they had a duty to these taxpayers that they needed to communicate more directly to the public about what they were seeing in their data. They started to reach out a lot more and make their science more accessible," he said.

"That's a good thing, right?"

"Yes it is. But some industries started to push back on this message, which made the scientists try even harder to talk about what they were seeing and their concern for the future of human civilization. So over time, the conversation moved from one of pure numbers about the topic to actually expressing how they felt about the future ability of the Earth to provide the resources like clean air and a livable temperature for humans to survive on the planet; their concern for the long-term sustainability of the human race was dire. They found that neither of those approaches worked to affect change in the public response to climate change," Per said.

"What's an example of this scientific message that doesn't work for the public?"

"Well, a great example is the new Hockey Stick Curve. You discussed in *Exposé on Climate Change* the controversy over the Mann, Bradley, and Hughes 1999 publication on the temperature reconstruction over the past thousand years. [2] A group of researchers led by Dr. Rob Wilson from the University of St. Andrews in Scotland and Dr. Kevin Anchukiatis at the University of Arizona published an updated Hockey Stick Curve. [3, 4] This is the state-of-the-art reconstruction including more thousand year-long, good quality tree-ring chronologies than the Mann, Bradley, and Hughes work. This new curve took special care to make sure that the low-frequency variability was preserved in the chronology so that it shows the Medieval Warm Period and the Little Ice Age as they have been reconstructed from many other records including historical documents. This research took many years and includes about 50 co-authors because of all of the work on all of the chronologies that went into the record. So this is very sound science. It shows the same basic trend of the one in 1999 with modern temperatures being higher than anything that the Earth experienced in the last 1100 years. Also, the rate and magnitude of increase in the last 100 years is greater than anything we have seen before in this record."

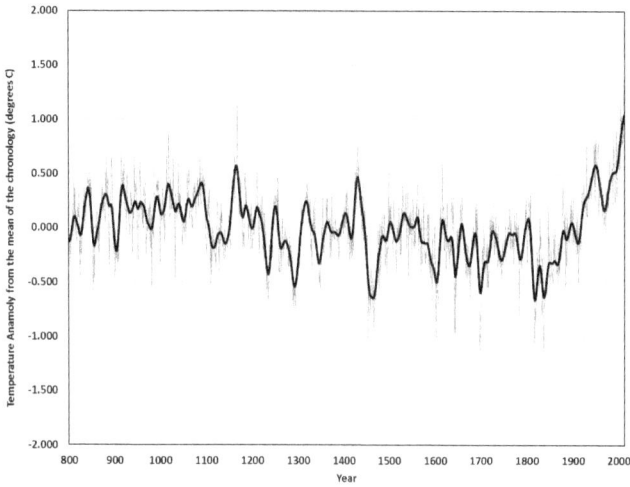

Temperature reconstruction from tree rings across the northern hemisphere. The temperatures are graphed as a temperature anomaly from the mean of the whole chronology. This work was published by Wilson et al. 2016. [3]

"That sounds very convincing."

"To a scientist it is. It's stronger data, done by a completely different group of scientists that have about 20 years of perspective on the Mann, Bradley, and Hughes paper. They have addressed all of the scientific critiques of the past studies and have reconstructed the most accurate record of temperature that we are capable of doing at this time. But it still is not enough to convince the climate denier or even most lay people," Per said.

"Why is that?"

"We have studied the psychology of this and have worked to identify where these messages fall short," Per said.

"This is your new research that you are just publishing now, correct? What are you finding?"

"We found that these more dire messages are no more helpful than the scientific speak which is too burdensome to understand. We find that people respond to these dire views of the future, which are likely to happen, with the five Ds. They are Distance, Doom, Dissonance, Denial, and Identity. [1] In other words, the first response is for people to say that this is a problem over there, far away. The seas are rising so those islands are affected but it won't affect me. Then the feeling of Doom often leads to an over response of disenfranchisement. The feeling that the individual themselves, can't do anything to make a change, so why try. Dissonance is where they see things changing on the landscape, but it does not match with the stories they tell themselves. This is often a psychologically painful state so people avoid listening to it or thinking about it. Then Denial is where they face the problem head on, but say that it is not happening because scientists are making up the data or are just environmentalists trying to get their way. Finally, people withdraw into Identity where they find like-minded people and put up barriers to hearing any counter view about climate change," Per said.

"That sounds bad. Is it even possible to reach the people in these final stages?"

"Well, to combat all of these stages we need to be using the five Ss, which are Social, Supportive, Simple, Signal, and Story. Social is connecting with peers and seeing what they have to say. It turns out that we all have a strong influence on our peer groups through Facebook and other social media. It is such a strong response that people end up unfriending those that don't agree with them, which creates smaller isolated groups that are not sharing ideas and adapting to change. But Social can lead to some change and swaying some people's point of view. Then Supportive is the idea that you need to provide three positive concepts for each negative one. So you can talk about how the temperatures are warming, a negative piece of information, but also say that the solar industry is providing more jobs than any individual fossil fuel industry. Also, that people are adapting to these changes, and finally that we can have a cleaner and happier future if we adopt clean energy and calm the rush of our maddening work pace. We have found that it helps to keep it Simple. Often the message from scientists is too abstract with a lot of numbers and discussion of a complex system. The system is complex, but not everyone is going to grasp that. If a message is too complicated, it is not the best approach to convince people of your point of view. Signal means to demonstrate the changes that are being made to show that progress is being made. This makes the change real be giving examples of technology that is working or social change that is taking hold. Finally, Story is the most important because it's the personal stories of peers that they trust that will finally convince people that we need to make a change. It all comes back to personal interactions of contacts that sway the argument," he said.

"That does sound like a better approach that could result in a more meaningful interaction."

"Try it with family and friends the next time you talk to a climate denier and see if you can find a story from what you've experienced that helps to connect to them and show what you are seeing."

"I will. I think that it's important that we figure out the right way to convey this message to the public," I said as I excused myself and went back to the conference sessions.

Chapter 23

The Chase

Virginia
December 15

The conference went on for a few more days. I had the chance to complete a few more interviews and took notes during all of the talks. The Mountain Lake Biological Station in Virginia was a seven-hour drive from Indianapolis, and it was nice to get so much information from a relatively short trip. With my Prius, it only took nine gallons of gasoline to get there. That was less than one tank of gas and a relatively small amount of CO_2 emissions. I was able to meet with many of the leaders in the field of sustainability in a short amount of time. It was a long week away from home though, and I was ready to head back to Indiana. I packed up my car with my roller bag and headed down the mountain.

As I passed the main lodge, I saw a black SUV quickly pull out of the parking lot behind me. I only noticed it because it accelerated so fast that it sprayed gravel on the other cars. As I wound my way down the mountain on switch back curves, I saw the SUV pull close behind me on the winding mountain road.

I tried speeding up so that he would back off and accept the pace that we were headed down the mountain, but he just stayed right on my bumper.

The road was twisting and turning and we were going down the mountain fast.

I looked in the rearview mirror and the SUV was only about a foot away from my back bumper.

I tried to remember my driving education lessons, but they didn't really deal with how to best drive down a curving mountain road at twice the speed limit when chased by a large vehicle.

I braked a bit before the turns, but then accelerated through them. I tried to stay in my lane, but we were going so fast that it was hard keeping my car in the lane.

I could feel my tires sliding as we made the sharp turns on the road. We were in a 40-mile per hour zone, but my speedometer said we were going 60 miles per hour.

I came into a tight left hand turn and felt the car slide to the outside. I saw sparks in my side view mirror as I rubbed against the protective guardrail that's supposed to slow a fall over the precipice down into the valley below. The black SUV stayed right on my tail.

I realized that this guy wasn't just in a hurry, but he was actually after me.

We made a few more turns and all I could see behind me was the grill of the SUV.

As I accelerated into the next curve, I crossed over the double yellow line in the middle of the road so that I could better handle the curves. Just then, I saw an oncoming old red Chevy truck in the other lane. I swerved back into my lane, just missing the oncoming truck. I could hear him blasting his horn as he continued up the mountain.

We were about half way down the mountain and I was thinking if I could just hold it together for a few more minutes, I'd be able to get out in the valley were the road flattened out. There I would be safe.

I saw big curves up ahead and aggressively drove through the curves to stay ahead of the SUV.

It seemed with each curve, I was getting closer to the guard rail and I occasionally felt branches hitting against my window as I got to close to the side of the road.

Up ahead I saw one last big curve and it looked like the road straightened out after that. I thought I was going to make it.

But I was wrong.

As I pushed into this last curve, the SUV accelerated and slammed into my driver's side back bumper just before I reached the guardrail. It sent me into a spin and my car flew off the side of the road. I felt a moment of weightlessness and then everything came to a sudden stop.

I was disoriented and shocked. I think I might have been unconscious for a little while. When I came to, I found myself leaning on the inflated airbag that came out of the steering wheel. I had a bloody nose and felt like I had been trampled by a team of horses.

I started to reach for the door handle and could feel the car rocking back and forth. Looking out the window, I saw that the car was half hanging off a cliff. I had hit a tree and my car was stuck on the tree, but partly hanging over a 50-foot drop into the valley below.

I carefully opened the car door. I could just barely step out on to the rock ledge.

I slowly worked myself out of the car, careful not to shake the car too much.

Finally, I was able to get my feet under me and scramble up the slope.

My Prius had half of its back bumper torn off on the driver's side and the front passenger side bumper was wrapped around an oak tree. Luckily, that seemed to be holding.

I walked the rest of the way up the slope and came out on the road. The SUV was long gone. I could see the black streaks of my tires where the SUV crashed into my car and the path of my car as it left the road and flew over the side.

Just then, the red Chevy truck that I almost ran off the road came back down the road and stopped next to where I was standing.

"Are you alright?" the driver asked as he put the truck in park, got out, and came around to where I was standing.

"I think so. Sorry for almost running you off the road back there."

"I saw that SUV was riding right on your tail. What's going on?" he asked.

"I'm not sure. He followed me from the Mountain Lake Lodge. He stayed on me all the way down the road until this final curve and then he crashed into my bumper sending my car over the edge," I said.

"It's a good thing that tree caught your car. Otherwise, you would have had a long fall."

"I know. I got lucky."

"Let me call police and I have a friend with a tow truck that can pull your car up from there. Hopefully it can be salvaged," he said as he stepped away to make some phone calls.

I took out my own phone and called Jenny.

"Hello?" Jenny answered.

"Hi Dear. I don't want you to be worried and everything is OK, but I was just in a car accident," I said quickly.

"What happened? Where are you?"

"I was just coming down from the conference and a black SUV chased me down the mountain. He was riding on my tail all the way down and then finally hit my bumper on the last curve forcing my car off the road. It was like he was hoping I would make a mistake and crash the car on my own. Then when it looked like I was going to make it safe to the valley bottom, he pushed me off this final curve," I said.

"You think it was deliberate?"

"Yeah, he stayed right on me, just about a foot away from my bumper the entire way down the mountain, and then finally he forced my car off the road, just before a retaining barrier. It seemed pretty well planned and executed. I was lucky that a tree stopped my fall."

"Who do you think it was?"

"I don't know, but I'm only working on the one story right now. It seems that they don't want me to finish that story."

"We had that guy who came by the house the other month and threatened you. Now this. They're pressuring you on this story, but it sounds like they just tried to kill you. I can't believe they would go that far!?!"Jenny said.

"This's definitely more than I expected. Someone is with me now and he's called the police. We're going to try to recover the car and will see what the police have to say. I'll call you back after I talk to the police and find a place to settle in and figure out how I can get home. Love you."

"I love you too. Be safe."

————————————————

It all hit me after I hung up with Jenny. My legs gave out and I slumped to the ground.

The driver of the red truck ran over to see if I was OK.

"I think I'm fine," I said. "It just all overwhelmed me all of a sudden. A couple of feet to the side, and I would be splattered in the valley bottom down below."

"Yeah, you're lucky. Why was that guy after you?"

"I'm writing a story about sustainability and I think one of the businesses I am investigating didn't want my story coming out."

"Still, running you off the road is extreme."

"I definitely agree."

"It looks like you hit your head and nose hard. Do you hurt anywhere else?" he asked.

"I have a headache, and my arms and legs both have a dull ache."

"We'll get you checked out once the police and the ambulance show up. We are a ways from the closest station in Pembroke, so it will take them a little time to get here."

"Thanks for the help. Why did you come back down here?"

"Well, when I first saw you coming over into my lane, I just thought you were some idiot that was distracted by his phone. Then I saw the SUV bearing down on you and I knew something wasn't right, so I turned around up the road and came down to see if everything was OK."

"Well, thanks for the help. Did you happen to get a look at the driver in the SUV?"

"Yeah, a bit. He had dark hair and a dark complexion. He looked strong, too, like he worked out a lot. I didn't notice anything distinctive on the SUV, but I imagine that your paint and that dent in the back of your car will be an exact match to one on a black SUV around here."

I could hear the sirens in the distance and we waited while a police car and ambulance pulled up.

Two paramedics jumped out of the ambulance and came over to check me out.

"Was anyone else in the vehicle?"

"No, it was just me."

"We heard this was a two-car collision. Is the other vehicle down there?"

"No, they sped away when I went down the hill."

They put me on a gurney and shined a light in my eyes as they poked and prodded my abdomen and limbs.

"Does this hurt?"

"Not sharp, but I ache all over."

"You might have a concussion and you're likely to have some severe bruising, but I don't think anything is broken. We will bring you to the hospital in a few minutes," one of the paramedics said.

The police officer stepped up at that point and took my statement while the paramedics called in to the hospital and continued to check my vital signs in the ambulance.

The tow truck arrived at that point. The police officer took a few pictures of the position of the car and the skid marks on the road and then gave the tow-truck driver permission to pull my car back up to the road.

It was loaded onto a flatbed tow truck. I overheard the tow truck driver telling the guy from the red truck that I was lucky to walk away from that accident.

With the account from the red truck driver, the police determined that it was an intentional act and they put out an all-points bulletin for the black SUV with a damaged front passenger side bumper, but with only that description, they didn't have much hope of finding the vehicle and the person that attacked me.

Chapter 24

Downhome Christmas

December through January
Indiana

After a ride to the hospital in the ambulance and being kept under surveillance for about three hours, they finally released me. I had bruised ribs, a strained left arm and thumb. They declared that I did not have a concussion but gave me a pain reliever for my headache.

They released me from the hospital and I took a taxi over to the garage where my car had been taken.

The garage attendant took one look at me and guessed which one was my vehicle.

"Well, I usually offer the good news and the bad news, but there isn't much good news. It looks like the SUV took off a good section of your back bumper and driver's side back wheel-well with the taillights. The driver and front passenger seat-mounted side airbags, driver knee airbag, and front and rear-side curtain airbags all fired. It will cost about $5,000 to replace all of those, but the worst part is that the engine block stopped the car from going over the cliff when it met that tree. That broke the engine block, so your car is pretty much totaled," the garage mechanic said.

"I still think it's a good thing that it met that tree, so I don't hold any grudges there. I guess I'll see about renting a car to get back home. Will you file this report with my insurance company so that we can start to work on the insurance claim?" I asked.

"Yep. That's automatic. We have your insurance information and we will send it to them. I can have our driver take you over to the car rental place."

"Thanks."

I picked up a one-way rental from Pembroke to Indianapolis and headed home. I had been away from Jenny and David for four days now and was ready to be home. I had been able to get better sleep during the conference than I had in the past month, but I was sure that Jenny would be exhausted when I got home. Also, after the attack, I just wanted curl up in my own home for a while.

The drive home was thankfully uneventful. I parked in our driveway and limped up to the house.

Jenny met me on the steps with David in her arms.

"I was so worried about you. You look terrible," she said falling into a hug with the baby squished between us.

"It was scary. That driver definitely had it out for me and planned his attack well. What keeps bothering me is that he seemed to want it to look like an accident and only hit me when I was about to get away."

"What's so important about your article that would cause them to threaten your life?"

"I keep thinking back to my meeting in the MallMart offices. Their security guy looked very similar to the people that I have been seeing around at ISU, who visited us at the house, and now the description of the driver. They were all dark-haired military looking men that are swarthy. I think that security company is a little more than just protecting the corporate offices. I need to look into them some more, and maybe get some security cameras around the house."

"For now, I'm just glad that you're home," Jenny said.

"Me too."

I just stayed with my family and tried to recover over the following days. I gave Jenny a break so that she could catch up on her lost sleep while I was gone.

We were planning for Christmas with our newborn. We decorated the house with lights and put up a Christmas tree. There was a local Christmas tree farm where we could cut our own tree. They were a family-run business and this crop allowed them to maintain their farm. The entire family came together during November and December to label the trees and help the customers clean and load their Christmas trees.

Trees were an interesting crop from an agricultural perspective because they did not mature and become ready for harvest until seven years from planting.

The Christmas tree farm kept each section in rotation where about half an acre was dedicated to Christmas trees in any year of the rotation. Once that plot was harvested, new trees were planted to grow for the following seven years.

We purchased LED Christmas lights to decorate our house. Since we lived in an apartment before, we didn't have much in the way of Christmas decorations.

While we were shopping for decorations we realized how much stuff is produced for the purpose of decorating our houses for just a month and then stored away for the intervening 11 months.

We quickly realized that baby presents and well-intentioned relatives were the hardest impact on the environment. Presents started to roll in for our one-month-old David. We quickly had to put out word that he was not ready for presents and that if people wanted to get him something, they could donate to a savings account that we set up in his name.

We had so many friends whose houses were overtaken by too many toys that quickly accumulated from the years of birthday and Christmas celebrations. At this age, David wasn't even aware of the holiday, let alone looking forward to toys.

We had been overwhelmed with the idea of all of the things that we needed for the baby at the beginning. But once we let our friends know that we were pregnant, we plugged into an entire network of friends with babies. They handed down everything from strollers, to baby carriers, a changing table, and even a crib. Once we started looking for second hand baby items and furniture, we found out that there was an informal economy of handed down items that could provide everything that you needed for a well-appointed babies room.

We felt good about this as we were keeping baby items out of the landfill and just didn't need to buy new items. It also saved us a huge amount of money.

Sustainability can be a series of small actions that help us move towards a better future. We don't have to use as much stuff and we can simplify our lives where we connect more with our friends, family, and community. Happiness is driven by interpersonal connections rather than more money or more work that is done. Once we realize that, we can move towards a more sustainable society that is cleaner and easier on the Earth.

We went over to Jenny's parents for Christmas day.

They had a large artificial tree in the front room that was decorated with 30 years of accumulated ornaments. The amount of lights on the tree was enough to heat the front room in the wintertime, although this winter was unusual for its weather. We had large temperature swings where we were having some 50-degree Fahrenheit days followed by below zero temperatures.

It's interesting to think about the options of an artificial tree versus a live tree for Christmas. The live tree is real and living, but we generally kill it for Christmas, use it for a short period of time, and then discard it. The artificial tree is made out of plastic, but can be used for many years. Along with that come storage issues and then final disposal issues when it is done. So there are trade-offs for each kind of tree. At least the live tree is a carbon sink for a period, but then if it becomes trash or is burned, it releases that carbon back into the atmosphere. In the end, that is probably better than creating and using more artificial plastic that will be around forever. I guess the best solution would be to buy a live potted Christmas tree that you could plant after the holiday. Then the tree is around for a longer time and can act as a carbon sink. These small decisions help to advance sustainability in our homes.

"Greg, are you OK from your accident?" Jenny's mom, Jessica asked.

"Yeah, I am feeling much better. Thanks for asking," I said. "A week around the house helped me recover some."

"Hey, I saw on the news about a youth group out of Oregon that was suing the federal government for not doing enough to slow climate change. It was called YOUTHvGOV, I think," Jessica said.

"I've heard about that court case. It is a group of 21 youth from Eugene, Oregon and James Hansen who used to be the director of the NASA Goddard Institute for Space Studies. The case is called Juliana V. U.S. and is currently before the courts. The lawsuit was filed in 2015 against the Obama administration and was continued against the Trump administration stating that US government is violating the youngest generation's rights to life, liberty, and property by the government's actions not protecting essential public trust resources [1]."

"Do you think they have a chance to make a change in the US government policy?" Jenny asked.

"I hope so. The Trump administration has filed suit multiple times to get the court case thrown out, but it has survived those attacks each time. They finally got a court date in October, but further counter suits from the Trump Administration has postponed the hearing," I said.

"It does sound like the kids have a good case when they argue that their futures will be hindered by current inaction from our government. It's just too bad that the administration is playing legal games rather than lettering the kids have their time in court," Jessica said.

"I agree. The delay tactics have kept it out of court for three years, during which time more damage is done to the climate. I fear that it is only court cases like this that will change the way the United States deals with these serious issues," I said.

"The US is definitely not showing leadership on these issues," Jenny said sadly. "And our son, David, will be growing up in a world that has a much different climate than we grew up in."

I got a call from my insurance company in early January.

"Mr. Cunningham? This is Julian at All American Insurance Company. How're you today?"

"I am doing well, thank you."

"We are processing your claim for the automobile accident that occurred on December 15th. Sorry for the delay in calling, our offices were closed for the Christmas holiday."

"No worries. Thanks for getting in touch."

"As you heard from the report at the garage, your car is totaled. We will be replacing your vehicle in the amount of $20,000."

"We just bought that car last year for $26,000. We won't be able to replace the car with that amount of payment."

"I'm afraid that's all that I'm authorized to offer. Cars depreciate very quickly once they leave the lot, so that's all that we can do."

"What about the medical bills?"

"That's handled through your medical insurance which is a different company."

"Yes, I talked to them the other day and with my 20% of the bills and deductibles, I'm paying about $5,000 for my short hospital visit. Now it's going to cost us over $6,000 to replace the same car as well."

"I'm sorry Mr. Cunningham, but that's the contract that we had with you and all that we can offer."

"I'm not blaming you, I'm just a little frustrated with our insurance system. The police declared that the accident was not my fault, but I'm still stuck with $11,000 in bills just to get back to where I was before the accident."

"I understand sir, do you have any other questions?"

"Actually, can I ask you a question on another topic?"

"I will do my best to give you as much information that I can, sir."

"What does your company think about climate change?"

"Climate change, sir?"

"Yes, you represent an insurance company. Does your company have policies or a stance about climate change?"

"We follow the National Association of Insurance Commissioners 2008 report. [2] This states that all companies should have a Disclosure of Climate Risk. [3] Let's see. I have a copy of it here on my computer. The recommendations are far reaching. This suggests changes in building codes and land-use planning related to flood events and sea-level rise. It discusses discounts and credits for using green building materials when rebuilding after a disaster. It even suggests that the government appoint a climate change czar, using those exact words," he said.

"Wow, and that was back in 2008?"

"Yes, these committees met and created these guidelines under the Bush administration."

"Would you say that the National Association of Insurance Commissioners is full of liberal, left-leaning hippies?"

"Ha, no. I don't think that would be an accurate description of the managers."

"So the insurance companies have been taking climate change seriously since at least 2008?"

"Yes, that's correct. Climate change adaptation, mitigation, and resilience have a large economic effect on our industry. We have been planning for climate change and taking it into all of our decisions for more than a decade. Why do you ask?"

"I'm writing a story on sustainability, and a large part of sustainability ties into the economic stability of companies. If they aren't economically sustainable, they can't keep the doors open. It seems that insurance companies have been planning for climate change and the economic impacts of that on their industry for much longer than politicians have been supporting the idea."

"In our industry, we need to plan for future threats to our insured customers and the industry has been aware of the threat from climate change for a long time. We have been following the Intergovernmental Panel on Climate Change reports since they came out in 1998. I am actually a little surprised that it took the Commissioners until 2008 to publish a formal guidance statement on it."

"Well, thanks for your information. That is an important piece to the puzzle of sustainability and planning for the future under our changing climate," I commented as I said goodbye.

Chapter 25

Legacy

Indiana Skyping to Los Angeles, California
February

I had heard about a suburban redevelopment project in Los Angeles where developers took a couple of fancy McMansions in a suburban neighborhood and converted them to two family homes. They also developed an edible garden in a park-like setting using permaculture in between the houses. I was Skyping in to interview the people that ran the area and get a tour of their complex.

"Hello, this is Greg," I said into my computer monitor. I was sitting in my living room in Robertsville. It was January, but the sun was streaming in through the windows and it was unseasonably warm. The trees had started to break buds in our yard. I was concerned that they would get damaged when the inevitable cold days came back. Exposé on Sustainability

"Hello, this is Janet, and Chris is with me here as well."

"Good to meet you. I read about the Legacy project in Los Angeles [1]. Can you tell me about it?"

"It started during the economic recession in 2008 to 2009. Housing prices bottomed out. People could not afford to stay in their houses and banks were foreclosing their loans. The collapse of the real estate market was so pervasive and long-term that housing prices became very cheap. The banks were desperate to sell these assets because they no longer had much value. We had been talking about finding a place where a few of our friends could purchase property, live closer together, and start to develop edible gardens that we could use to feed ourselves," Janet said.

"We had some friends that were in real estate and they heard about these houses in a relatively rich suburb of Los Angeles. Each home had property and the neighborhood was located near to a public transportation project called PLACE. We knew that the city was working towards locating a light rail station near to this neighborhood and we realized that it had everything that we were looking for. It was a bit unconventional but we decided to aggregate our resources between four families. We bought two of the houses next to each other and split them into four living units for the four families. Then we converted the grounds using permaculture techniques to plant fruit and nut bearing trees with edible plants as the understory," Chris said.

"I imagine that you all stood out from the neighbors."
"We were definitely the odd folks on the block, but really a lot of these houses were abandoned and the remaining

people in the neighborhood were happy to see people moving in to the area and making a living here. Some of them said that our efforts were keeping the neighborhood alive," Janet said.

"Is it possible to get a tour of your place?" I asked.

"Sure, let me unplug the power from the laptop and we'll walk you around. One of the first things that we did was to install wifi throughout the grounds with boosters so that we could even work on our laptops from the garden. We spend a lot of time outdoors. Being California, it's comfortable outside for much of the year and we prefer to be outside," Janet said.

"We had way more space than we really needed for a single family in each house. The houses were over 3,000 square feet each, so we divided them into two separate wings with a shared kitchen. It only took a little renovation so that each family had its own living room and, of course, bedrooms. The shared kitchen was already in good shape, as people seem to invest in their kitchens, so that stayed the same," Chris said.

"I can see that it's a very modern kitchen that opens on a communal space. You must be able to entertain a pretty large group in there," I said.

"We have the other families over at least once a week for a group meal and then they host us on another day during the week. That helps us share the labor of food preparation. We also share childcare between the families so that the kids have their friends to play with and a parent can always be with them while the other grown-ups work," Janet said.

"Sometimes that work is away from the home, but other times, that just means being able to spend some concerted time in the gardens maintaining the plants or harvesting the crops," Chris said.

"We have a large back patio that goes straight out into the gardens," Janet said as she exited the sliding glass door into a lush garden in what used to be the grassy yard between the two houses. "By combining the yards and removing the grass, we have been able to develop over an acre of permaculture garden that comes right up to our houses. We have the more intensive crops, such as lettuce, tomatoes, and carrots up close to the houses." "I can see the one-foot square planting beds over there. It looks like you have a lot of food producing plants packed into a small space. Are those herbs in those raised beds?" I asked.

"We keep a wide variety of herbs such as oregano, chives, sage, dill, marjoram, and thyme. They are all very easy to maintain, as they are perennials. Since they keep growing year after year, we always have fresh herbs ready at hand for our cooking. The other vegetables are more seasonal, but we plant varieties and use cold frames to extend the season. We can grow some sort of produce for much of the year," Chris said.

"As we walk back along the path, we are getting into phase II of the gardens which are less intensively managed. You can see we grow apricots, almonds, apples, and pecans. These took a while to establish, but now that they are mature, they produce nuts and fruit every year and we just need to come out and collect when it's in season. The understory in here consists of raspberries, gooseberries, and strawberries. They can handle the shade and thrive in the dappled light underneath the lower crowns of the fruit trees," Janet said.

"Wow! That's a lot of food growing in a tiny area. You have many layers of plants from ground cover, to low shrubs, and even trees that are producing food for you. I can imagine just walking your paths and browsing on the trees for lunch. That's amazing," I said.

"It's one of the best parts about living here. California has such a good climate year-round for plant growth and a wide variety of food plants can be grown here. But you could do this type of food production in just about every part of the world. You just have to find what plants grow well in your region and then layer the vegetation so that every square foot is producing food on multiple levels. It makes for very efficient food production," Chris said.

"The key to it all is in the soil. You have to work with the ecology to support a healthy ecological system in the soils. Our process is organic where we don't use any artificial herbicides, pesticides, or even fertilizers. We encourage helpful bacteria and insects in the soil. We compost all of our scraps and actually collect leaves from our neighbors that we compost as well and use as a soil additive," Janet said.

"There's a horse farm just down the road and they needed a way to get rid of the manure. It turns out that horse manure is nutrient rich and a very good organic additive to soils. We get a truckload of their manure every few months and spread it under the trees. That helps keep the soils healthy and replaces the necessary elements that we are removing as we harvest the fruit and plant material," Chris said.

Chris and Janet continued to walk down the winding paths where all that I could see was a tunnel of vegetation that was surrounded by fruit from the berry bushes on either side with fruit hanging down above the trail from the taller trees. The path meandered on further and started to get sparser where I did not see as much evident fruit.

"This is zone III of the gardens. Out here are oak trees that produce acorns and native poplar trees. The understory is full of bee balm, Rudbeckia, and other native plants. This zone is more for the wildlife. Birds and deer come into this zone and find good food. That way they have their own food here and tend to leave our crops alone that are closer to the houses. It encourages pollinators to make the area their home, which helps our fruit trees and bushes as well," Janet said as we walked along a small creek that used to be a drainage around the houses. There was a little bit of water in the creek and I could see that this would keep everything well-watered.

"I like the creek. That's a nice amenity."

"That was one of the reasons that we chose these homes. The creek was considered a nuisance by the previous homeowners because their property sloped down to it and they could not keep as much grass. It also flooded on occasion during the rainy season when the whole area was cleared and everyone had just grass in their yards. Now with the trees that we planted in here, they take water from the creek and we have not had any flooding in the past five years. It helps to keep everything watered and encourages the wildlife to use this part of the property," Janet said.

We continued to walk, the trail headed up to the other house. I could now easily see the different zones as the vegetation changed back from the higher trees with the sparser understory to the shorter fruit and nut trees and the rich understory of berries. Finally, we came up to the back porch of the other house and their own kitchen garden at their back patio.

"This produce garden mimics our own, although they have chosen a different selection of produce. We all maintain all of these spaces and then we have the variety of produce from the two gardens and the permaculture canopy forest. Also, with the variety of plants that we have, there is usually something producing throughout much of the year," Chris said.

"This must have taken a lot of planning to figure out what would be best to grow together and could tolerate the shade. The zoning must also takes a lot of planning," I said.

"My mother worked most of her life in studying native plants and edible crops that would do well in this climate and worked this plan out over about 30 years of experimentation on her own place. We were able to use her knowledge to develop our permaculture area," Janet said.

"Well, thanks for the tour. It was great to see what you're doing there. I'm glad we were able to use Skype; it allowed me to be able to enjoy our meeting from the comfort of my own home."

"We appreciate that you can learn from what we've done here, but don't have to burn the carbon to fly out to see it. This is the best way to give a tour and pass along what we have learned. We actually do about one of these a month and have virtual visitors from all over the world. It turns out that there's an active agritourism industry that we can tap into. If you would like to log into our website and leave a donation, we'd appreciate that. Any funds that we raise are used for our educational outreach mission," Janet said.

"This is a great resource that we can share with others. We're able to produce more edible food and a greater variety than any industrial agricultural process. We do it with much fewer inputs and much less cost. We have mainly chosen perennials, so instead of depleting the soils and having to clear crops every year, it just gets stronger and stronger as the years go by," Chris said.

"Thanks again. I'll definitely leave a donation and will be in touch if I have any questions," I said as I closed out the Skype call.

Chapter 26

The Very Air We Breathe

Wabash River, Indiana

March

I met with a few friends to paddle on the Wabash River near Terre Haute, IN. Dr. Seamus Flanagan came along and brought Dr. Tom Sanford a Sociologist and Dean Jack Murphy.

We each had our own kayaks and enjoyed being on the water, but none of us had enough time to regularly get out on the river. We were happy if we could paddle once or twice a year. We decided to put in at Fairbanks Park, paddle upstream past the Landing, and up to the Wabash River Generating Station. After that, we would float back down the Wabash to our put-in location. This made transportation easier since we did not need separate drop-off and pick-up points.

March was a tricky time to be on the Wabash because the current could be faster due to floods and the water could be cold. It had not rained much lately so the current was not too bad. The whole round trip would be about 14 miles of paddling and take the whole day.

We pushed off early in the morning; the river was foggy from steam rising from the relatively warmer water into the cool morning air. We eased our kayaks into the placid water; it was a cold March morning but a beautiful scene before us. Most of the trees were still bare, but a few were leafing out, providing a light yellow-green color to the landscape. Rich brown colors painted the surrounding hills from last season's leaves. A blue heron flew up from the shoreline as we paddled up stream.

Our yellow and red kayaks added bright splashes of color to the muted scenery as we paddled up the stream.

The single person kayaks made good time in the water and we paddled in easy cadence.

"So, Seamus, what's the brown color in the water from?" I asked.

"Many people are afraid of the waters in the Wabash, thinking that they are polluted and hazardous. But most of the color comes from diatoms, which are a golden brown algae. These organisms create internal shells, called tests out of silica and those forms continue to float downstream after the algae die. Some of the color comes from agricultural run-off, which includes silt and clay, but most of the color is from the diatoms," he said.

"It's interesting that people in Terre Haute are concerned about the water in the Wabash," Tom said. "In the 1960s, people used to have a boat parade on the river. They would put in at the boat launch at Fairbanks Park in late spring and motor up stream. They would picnic on the sandbars and that would open up the recreation season on the Wabash. And that was before the Clean Water Act that regulated industrial discharge into the river, so the water quality was probably much worse back then."

"And before that," Jack said, "people used the Wabash as a major transportation corridor. All the way back to the Native Americans and the first European settlers, the river was used for transportation. That's why Terre Haute is located where it is and where it gets its name. Terre Haute means high land in French and the earliest French fur trappers used the area. Through all of its history until recently, the river was used for transportation. There are some historical photos of paddle wheel river boats on the Wabash taking people out for recreation."

"The arts community is trying to bring appreciation of the Wabash back to public awareness. An organization called ArtSpaces declared 2013 as the Year of the River. They had over 100 organizations hold events related to the Wabash River in that year. ArtSpaces is working now to acquire grants to help develop a walking path from the historic Wabash road, past the courthouse, and to the edge of the river. They hope to create an outdoor stage on the banks of the Wabash that will provide another concert venue and outdoor gathering space. This will improve walkability in town by creating walking trails along the Wabash connecting Fairbanks Park through Indiana State University property to the Heritage trail on the north side of campus," Tom said.

"The Heritage trail was a Rails-to-Trails Program that was led by Pat Marvin. That trail extends almost seven miles and runs from ISU to Rose Hulman Institute of Technology and beyond. More walking and biking trails connecting to that will help improve human powered transportation around town," Seamus said.

"The Wabash River itself is a water trail with a canoe route and campgrounds all along its 500 mile length from Ohio, across Indiana, and joining with the Ohio River in the southwest corner of the state of Indiana. We are starting to see many more outdoor recreation opportunities which improves human health and happiness," Jack said as we paddled up river and crossed beneath the Hwy 40 bridges and the train trestle.

"In Sociology, we define that as 'Quality of Life'. It's an interesting parameter to measure because whatever drives one's quality of life differs for each person," Tom said.

"Yes, but people know when they have a better quality of life even though sometimes they can't point to what makes it better," Jack said.

We paddled around the bend where Indiana American Water Company was located that pumped water from the underground aquifer and treated the water that the City of Terre Haute used.

"That's where we get our water from. They draw it up from the aquifer, although many people think that it comes from the Wabash since the fresh water plant is next to the river. It's a good thing that we are not taking the water directly from the Wabash because the process of the water percolating through the soil layers cleans it of many contaminants," Jack said.

"Even though the brown color in the waters mainly comes from diatoms, the Wabash River waters are not completely clean," Seamus said. "The Indiana Department of Environmental Management regularly tests the water quality at stations since the late 1990s. I did some research with an undergraduate student to examine all of this data for the Wabash River. We looked at how the river draws down heavy metal contaminants, and also examined the main industrial point sources of these heavy metals. The Wabash River Generating Station we see ahead of us produced about two million pounds of EPA regulated elements or chemicals every year based on the EPA Toxic Release Inventory data. It creates about 98% of Vigo County's pollution burden. The generating stations burn coal to produce electricity for us and put out large quantities of lead, arsenic, and mercury as a biproduct."

"Where do those pollutants go?" I asked.

"Some of it is out the smokestack although scrubbers were mandated by the Clean Air Act in the 1970s to remove the worst of the air born pollutants. Still some of those heavy metals escape into the surrounding atmosphere. Some of them are captured as fly ash coming out of the smokestacks and dumped into coal ash ponds or in landfills. There are some coal ash ponds just about a football field distance from the river we are paddling," Seamus said.

"What did you find when you analyzed the heavy metal levels in the Wabash River?" Jack asked.

"The Criterion Continuous Concentrations are posted by the Environmental Protection Agency for the level of any pollutant that would damage organisms if continually at that level [2]. We found that lead, arsenic, and zinc were frequently above the Criterion levels. Multiple stations along the Wabash River showed spikes in lead concentration in many years but 2008 and 2009 were extremely hazardous. We also found that these heavy metals are transported 50 to 100 miles downstream before levels returned to normal. And this means that the heavy metals are incorporated in the organisms and sediment in that 100 miles. We are definitely polluting too much for the health of fish, birds, and insects in Indiana. Also if you eat the fish or even the animals that you hunt in this area, and most of the eastern US, you will be consuming some of those heavy metals," Seamus said.

"Why is that? Don't the elements settle out and go away?" I asked.

"There is no "away". Even if the heavy metals go into the sediment, which is the best-case scenario, bottom-feeding organisms root around in that sediment and consume some of the heavy metals. Then other organisms eat those and 100% of the heavy metals are passed on to the organisms that eat them. This is called bioaccumulation. Natural organisms don't have any process to change or

secrete these heavy metals or the chemicals that we produce, so they are passed up the food chain. Then when
we eat from that food chain, we eat those contaminants as
well," Seamus said.

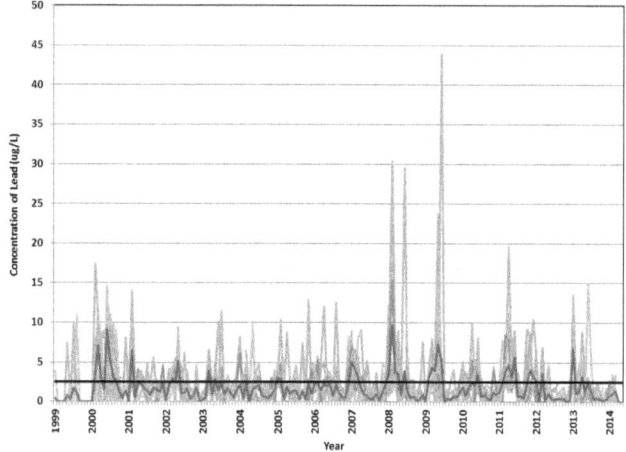

Lead concentration in the Wabash River from 1999 to 2014 [1]. The
black horizontal line is the EPA established Criterion Continuous Concentrations where organisms start to be poorly affected when levels are
above that limit. [2]

"That's bad for the natural system, but it doesn't affect the beauty of the river," Tom said, as we paddled
around another bend and got a long view up the river. We
could see ice floating down the river towards us.

"I saw a hand-drawn postcard at the Clabber Girl mu-
seum from the 1800s that showed people cutting ice on
the Wabash in Terre Haute during the Little Ice Age in the
mid 1800s. The weather doesn't get cold enough to freeze
the Wabash at Terre Haute anymore, so it would not be

possible to do ice cutting today. But it's neat to see the ice floating down from the colder north. This really only happens at this time of the year, as the waters increase in strength and wash ice down from further upstream," Dean Murphy said.

"This is a wonderful natural area and the Wabashiki Fish and Wildlife Area is just across the river from Terre Haute. That is a protected Fish and Wildlife Area that covers more than 2,500 acres of land. We had a master's student study the birds out there and she documented more than 100 species of birds. In one afternoon in the fall, you can see hundreds of great blue herons and great egrets on the ponds at Dewey Point. They feed on the fish in the ponds from when the flood waters rise on the Wabash and wash fish into the wetlands. The area seasonally dries every year, so that the ponds get isolated and it's easy fishing for the birds. This is developing into one of the more important birding areas in Indiana and the Midwest as a whole. It's a great resource for the area," Seamus said.

We continued to paddle up the river and took a break at the Landing, which was an old landing point for boats. There was a restaurant and events center just up the bank from where we pulled our kayaks onto the shore.

"This is near where the original Fort Harrison was built in 1811," Tom said. [3] "Just up on the bank there was the first European fort in this area. That is the reason that Terre Haute was located here."

We unpacked our lunch of cheese, sausage, and bread that we shared while we drank water from our metal water bottles.

"It's depressing that most people ignore the Native American history of an area. We usually start counting history when Europeans arrive, but that's relatively recently compared to the Native Americans," Seamus said. "There is still some debate about when and how Native Americans arrived in the Americas, but most estimates put it at about 15,000 years ago. They could have walked on land through a gap in the melting Laurentide Ice Sheet or paddled along the coast. The latter was more likely, but that also means that any evidence of their early habitation sites would be about 150 feet below sea level. As the glaciers melted, sea level has continued to rise and any shoreline camps would be under water."

We finished our lunch and got back in our kayaks.

As we paddled north, the land became flatter and agricultural fields spread out on either side of us. More ice was in the water, but it was easy enough to avoid.

We could see the smokestacks in the distance and it was not too long before we came up to the generating station. The building loomed over us and water was being discharged into the Wabash. We could feel the water warming underneath us from the discharge from the generating station. White smoke roiled out of the smokestacks some hundred feet above us.

"This generating station is slated to shut down. They lost a court battle for non-compliance with the amount of air pollution they were generating. That will clean up the local air and improve the contaminant burden in this county, but it means that we are now getting our energy from the Cayuga Generating station. That is further away and actually produces four times the amount of pollution because it's a larger generating station," Dean Murphy said.

"Closing the plant will put some people out of jobs. This plant has a state-of-the-art coal gasification facility. I believe that will keep running when they shut down the five boilers that were producing electricity for this region," Tom said.

"It's a complex issue," I said. "We all use electricity so we need to source it from somewhere. But it would be better to reduce our use and conserve as much energy as possible. Also, we could switch to alternative energy such as solar, wind, or geothermal energy. The difficulty comes when fossil fuel industries insist on sticking with the processes they have always used instead of investing in new technologies to provide cleaner energy. Our government could provide subsidies and make laws that speed up and improve the transition to alternative energy, but the fossil fuel industries are the ones that currently have the money and power. And they are using those resources to hold on to as much of the industry they can instead of being innovative and leading the change to alternative energy."

"Look up ahead there. Do you see the orange water coming out of that tributary? Let's go check that out," Tom said.

"This doesn't look good. Bright orange water flowing into the Wabash. What is this?" Dean Murphy asked.

"That is called acid mine drainage," Seamus said. "This is near the Green Valley Mine. They used to mine coal back there and they didn't remediate the landscape after mining. As rainwater washes over the tailings piles of rock left over after mining, it breaks downs minerals like pyrite and produces the metal-rich effluent that is orange in color. It's full of iron and lead. It's not from the generating station but from a different company's property that was mining coal that was used in the generating station."

"Is there any way to clean it up?" Tom asked.

"There is a micro-organism that one of my colleagues in the department discovered. It's called *Euglena* and it takes in the iron and makes iron nodules. As the organisms die, it precipitates out the iron reducing the acidity of the water and cleaning up the acid mine drainage. They hope to be able to use this to remediate these sites in the future," Seamus said.

We turned our kayaks around at that point and started paddling back down stream. We had been paddling up stream against a gentle current. On our way home, we floated with the current, but an upstream wind actually pushed against us more than the current pulled. So it was still a hard paddle. It was late in the afternoon when we arrived back at Fairbanks Park and took our kayaks out of the water.

"That was a good paddle," Dean Murphy said.

"We covered 14 miles in six hours. Not a bad day," Tom said.

"It's amazing to find such a wonderful recreation area in your own backyard. Not many people use the Wabash River for recreation these days, but it's a great resource and should be used more often. People just have to get over their perception that the river is too polluted to use," Seamus said.

Chapter 27

Chasing Happiness

Bhutan
April

Bhutan uses a happiness scale when polling its populace to determine if the government is doing the right thing. I wanted to visit this unique Asian county in the Himalayas, but it is hard to get to. I couldn't book a direct flight to Paro in Bhutan so I flew into Kathmandu and then booked a connecting flight into Bhutan on Drukair. Druk means "thunder dragon" which is fitting for the mountain kingdom of Bhutan.

Getting a visa for Bhutan was not easy either. I had to convince my contacts in Bhutan that it would be worthwhile giving me an interview and then they had to invite me into the country to meet with them. Visitors from outside the country are also restricted to visiting just two valleys. Paro has the main airport and Thimpu is the capitol city. Outside of those regions, visitors are not allowed to travel without special permission. They want to maintain their culture and identity and do that by restricting visitors and travel within the country.

On the flight from Kathmandu, I was lucky enough to have an eastward facing seat and the pilot announced that we were flying by Mount Everest. I had spent some time in Nepal before and never got to see Mount Everest from the ground. It was incredible to see it out the airplane window with a white puffy base of clouds and this massive mountain protruding above them. I had never wanted to hike Mount Everest, but looking out at its cold peak covered in rock and snow filled me with a peaceful feeling.

We all marveled at the countryside as we approached the Paro airport. We were in a large Airbus A319, which had six seats across that held 114 people. The pilot manually maneuvered the plane through the hills of Bhutan into the 7,300-foot elevation airport that is considered one of the top ten most dangerous airports in the world. As the pilot maneuvered the plane, we could look down on the lush tree-covered hills at the pine and fir forests. Sunlight came streaming through the clouds, highlighting villages and mountains out the window.

When I landed and exited the airport with my bags, I met Dorji Tenzin. He is the deputy secretary for the Gross National Happiness Commission in Bhutan and had agreed to be my guide and mentor while I was in the country. He was a compact, strong man, with his hair cut short. He looked like he could have had a military background and was smiling in a friendly way.

"You must be Greg Cunningham," he said as I walked up to him as he stood holding a sign that read Cunningham in the mass of people at the airport.

"Yes, and you are Dorji?" I asked.

"Glad to meet you. Our car is over this way. We have about a one hour drive to Thimphu," he said as he led me to the parking lot.

He put my bags in a small four-passenger car. It was so compact, it was hard to fit my rollerbag into the back.

We sat in the back seat and had a driver who negotiated the road out of the airport.

"Your English is very good," I said.

"English was made the second official language in Bhutan and it is taught in all of the schools. So the younger generation can speak English quite well. It's mainly our elderly that never learned English. As the second official language, even our official signs are written in English as well as Dzongkha," Dorji said.

"That must make it easy for westerners like myself to travel in Bhutan," I said.

"We have many visitors from the west and they seem to function quite well in our country. I understand that you are here to learn about our concept of Gross National Happiness?"

"Yes, I'm curious about alternative ways of measuring success for a country," I said.

"As you know, most countries measure wealth by the Gross Domestic Product, which is a measure of the goods and services developed from a country, but there are some problems with that as a measure. Gross Domestic Product measures money that is spent whether it's on a good thing or a bad thing. So the massive BP oil spill in 2010, in your country, added to your Gross Domestic Product because it cost billions of dollars to clean it up. But if a person volunteers their time to help clean up the oil spill, that is not counted in the Gross Domestic Product even though their work is commendable. The Genuine Progress Indicator is a better measure that takes the Gross Domestic Product, subtracts money spent on environmental damage and other negative impacts, and then adds the value of things like volunteerism. This is a more true measure of the health of the economy and well-being in society."

"I was looking into the Gross Domestic Product in the US," I said. "When you plot the Gross Domestic Product for the United States, it is heading steeply up, but if you look at the Genuine Progress Indicator, it has flattened out for the last 20-30 years, which is more reflective of how the general population feels. Many countries don't want to use the Genuine Progress Indicator for this reason. It doesn't show their country in a good light, but it's more representative of the state of the country and quality of life."

"Economists have suggested changing to the Genuine Progress Indicator a while ago, but it's not catching on very quickly. More countries need to make the Genuine Progress Indicator the standard measure of economic well-being. It is a more realistic measure of how people feel. Bhutan's approach of measuring happiness is another good approach," Dorji said.

"Tell me more about your measure of Gross National Happiness." I said.

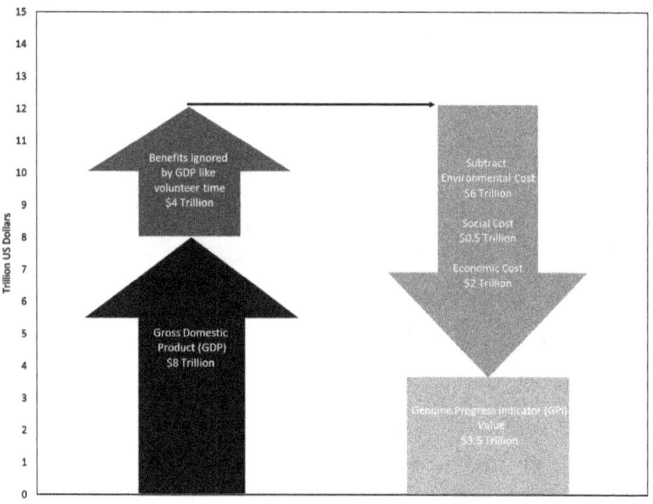

The calculation of the Genuine Progress Indicator starts with the Gross Domestic Product, then benefits ignored by GDP, such as volunteerism is added to that valey. Negative effects such as environmental costs, social costs, and other economic costs are subtracted from that total providing the value of the Genuine Progress Indicator.

"In 1972, our fourth king suggested the use of Gross National Happiness, stating that 'Gross National Happiness is more important than Gross National Product'," he said. "But then we needed to figure out how to measure it. Currently we send out a lengthy survey asking people about their happiness. Some of these surveys are 40 pages long."

"It seems that happiness is harder to measure than income."

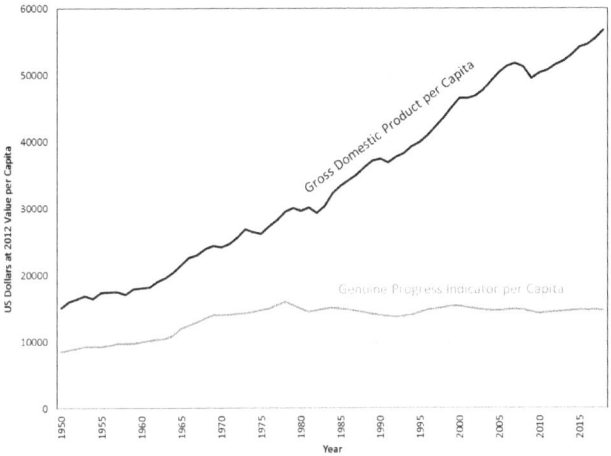

Gross Domestic Product versus Genuine Progress Indicator from 1950 through 2018. GDP data is from the Federal Reserve Bank of St. Louis and the GPI data is extrapolated from multiple graphs online.

"It is hard to measure, and figuring out why people are happy or not is even harder. Much of the survey deals with quality of life questions: are people fulfilled at work, getting enough exercise, getting good quality food, do they have good community connections, and more. There are many aspects to what makes people happy and our government works to provide the basic needs so that people can be happy in our country," Dorji said.

"I went to conference last year in Nepal and interacted with two scientists from Bhutan that were doing tree-ring research. Even they were talking about happiness to the other participants at the conference. Your citizens are really pushing out this concept of happiness to others and it has become a hallmark of the country of Bhutan," I said.

"The people really appreciate the fourth king who is now retired. They respect the idea of measuring happiness and the idea that Bhutan originated this idea that is affecting the global discussion."

I gazed out the window as we drove to my hotel. "These mountains are amazing. The valleys are so steep and the road is right at the edge. I can see the river flowing down below," I said.

"We have great natural resources in our country. Most of the country is in the Himalayas so we have many streams and good water resources. One of our major exports is hydropower that India buys from us. This allows us to have energy that is clean, leading us to be the only carbon neutral country in the world," he said.

"It sounds like the people of Bhutan really value their natural resources."

"Economists call this natural capital and we do truly value it in our country. It is interesting that economists have tried to value other things besides marketable goods and services. There are two fields of economics that have examined the value of natural capital called Environmental Economics and Ecological Economics," Dorji said.

"They both sound like they would have a lot of concern about the environment."

"They sound that way, but Environmental Economics relies fully on neoclassical economics. They believe that the markets will be able to adequately value natural resources and they also assume that if a natural resource is used up, then a substitute for that resource can be found. Ecological Economics takes a completely new perspective and approaches their understanding of nature and economics from the point of view that resources can be used up and *not* replaced with other manufactured products. Ecological Economics values the natural system for itself. One example that they provide is the ozone layer in the Stratosphere - the second layer of the atmosphere above where most planes fly. Ecological Economists rightly point out that we have not found any way to replace this layer, so it is finite and cannot be replaced by another commodity," he said.

"That does sound like Ecological Economists disagree with some of the basic tenants of neoclassical economics."

"This is an important distinction because it starts to accurately value the natural system and not just see it as a source of goods and services that are external to the economic pricing of objects. It provides true value to these ecosystem services and the products that we get from nature with the understanding that if we use them up, they may be gone forever."

"That is an important change in perspective. It takes a more ecological viewpoint, realizing the balance in the nature and the value of what we glean from nature. Are many economists taking this viewpoint?" I asked.

"Sadly, the Ecological Economy view is still in the minority, but I think as our resources diminish and we feel more pressure from changes in the system through issues like anthropogenic climate change, more economists will start to value nature as they see their classical economic models start to fail."

"Not to change the subject, but I love the architecture of all of the houses and businesses throughout this area," I said as we started to enter into the outskirts of Thimphu.

"Our government outlined building codes in 1998 that preserved our traditional construction techniques and the appearances of our buildings, so even brand new construction in the heart of the city has the traditional look of Bhutanese buildings," Dorji said.

"I really like it. There is so much wood in the structure and many of the buildings have paintings of dragons and other things on the walls."

"The paintings are mostly of the dragon, snow lion, tiger, and the garuda which is a mythical bird. They represent elegance, purity, strength, and fearlessness. You will also see paintings of the four harmonious friends, which are an elephant, a monkey, a hare, and a bird underneath a fruit tree. There are various stories about the four friends, but most of them revolve around the need of the four animals to help each other for them all to flourish. This is an old tale from Bhutan and Tibet and I believe that it relates the origins of our conservation ethic."

At this point, we pulled up to the Khang Residency in the upper part of Thimpu. The driver brought my bags inside as I checked in at the front desk and Dorji said that he would wait in the lobby as I got settled in.

I walked the stairs up to my fourth floor room and was surprised to find that it was a suite with a large living room, a kitchen, and a separate equally large bedroom with a bathroom off it. I quickly dropped my bags and cleaned up from my long air travel. I took a minute to step out on the balcony and look at the view. My room overlooked most of Thimpu from a bit up the mountainside. I could see all the way across the valley and the expanse of the city. It was a beautiful city with prayer flags of multiple colors fluttering in the breeze on the mountainside. There were areas of pure white flags in different locations around the town. A storm had just blown though and dense white clouds clung to the valley bottoms while wispy tendrils of clouds played about the surrounding mountains. The city itself sat in a bowl with mountains all around it. The setting was truly marvelous with a peaceful feeling to it.

I went down stairs and joined Dorji who was standing in the front lobby.

"Good morning. Are we taking the car?" I asked.

"Our car was a rented taxi that I took from Thimphu to Paro to pick you up and of course, brought us back here. I paid the taxi driver and thanked him for his service. We can walk from here," he said.

We left the front of the building and walked to the right, slightly downhill as we headed towards the Royal University of Bhutan and the government office for Gross National Happiness. After about ten minutes of walking first down a street and then up a fairly steep road, we came to the university. There were paved sidewalks for part of the way, but they were not always connected throughout the area so we occasionally walked in the street. I found Bhutan to be much quieter and the drivers much more polite than in Nepal. Cars seemed to follow standard driving rules and made way for each other. The roads were not very crowded and cars made way for pedestrians.

We walked through an arched gate into a flagstone courtyard. The buildings around us housed large presentation halls and classrooms, but still had the similar wooden structure and traditional paintings on the walls. We made our way to a small meeting room that had bookshelves on all walls and a large window overlooking the courtyard.

"This is the center for Gross National Happiness in Bhutan. We are a small group and spend much of our time working on surveys with the public to get a measure of the happiness level from many corners of the country," Dorji said.

"Is it difficult to travel to around Bhutan? There are a lot of steep mountains and remote locations," I observed just from the drive between the main two towns in the country.

"This is true. There are a lot of remote provinces. But we find that the people throughout the country are the same. They are willing to take the time and sit through our questions because they think happiness is an important measure."

"You mentioned that this was brought up as a metric by the last king. Does the current king support this as a measure of the country's wellbeing?"

"He certainly does. The fifth king is the son of the king who put this idea in place and he fully embraces the concept of happiness," Dorji said.

"What can the government due to promote happiness?" I asked.

"They can make sure that basic needs are provided. Everyone gets a good education, has health care, that the parents have work for an income, and that they can support their family. Once those basic needs are met, we find that people need community and connection with each other. That usually comes up on its own in the small villages. But the basic needs are the domain of the government."

"In the United States, so much of our idea of wellbeing is tied to our income and status. It seems that the United States is moving further from providing those basic necessities you listed and we have never had universal healthcare for all of our citizens," I said.

"That is one of the important basic needs. If your people are not healthy, they can't work and contribute, then they don't have money to support their families, and the entire economic system degrades.""I completely agree, but so many people feel that it needs to be something like a free market system and that people need to be able to purchase the healthcare that they want. The only problem is the main people saying that are the rich, that don't seem to understand that many people can't afford basic healthcare in our own country," I said.

"You know, our system is not the only one that measures happiness. The Happy Planet Index measures Happy Life Years that are calculated by combining a country's wellbeing, life expectancy, and inequality outcomes combined into a single measure [1]. This is compared to the ecological footprint that a country uses to reach that level of Happy Life Years. By these metrics, Costa Rica is often at the top of the list of the highest Happy Life Years at the lowest environmental cost. Luxembourg, for example, has a similar Happy Life Years score, but at a huge environmental cost."

"It is interesting to compare happiness to the ecological footprint of a country. Where does the US fit into this scheme," I asked.

"The United States is in between these two extremes, but still in the red zone of the Happy Planet Index because of its ecological footprint."

"How is it that Cost Rica scores so high?"

"Costa Rica has a low ecological footprint through alternative energy, disbanding its military so that it has less negative costs on warfare and can do more to support its populace, and then a large portion of its economy is driven by ecotourism. This latter metric is a feedback with environmental quality because it depends upon a healthy environment to drive tourism.

We continued our conversation over the next few days and I got a tour of the university, the government offices, and the city. Once my interviews were done I was excited to squeeze in a visit to the Tiger's Nest, a remote monastery perched on the side of a cliff, before I returned to the United States.

Chapter 28

The Trap

St. Louis, Missouri
May 11

I knew that MallMart and the swarthy security company operatives were following me and I was getting tired of being chased. It was time to poke the bear.

I made a few arrangements and travelled to St. Louis where the MallMart CEO Roy Anchorage was speaking at an industry convention. I was able to get press access to be behind-the-scenes and have interview access to many of the speakers at the conference.

MallMart was going to be making a big announcement that day about how all of their stores would only use LED lightbulbs and how they would double their organic offerings in all of their stores across the nation. There was a press conference before the main speech, so I found my way back to the pressroom.

I saw Mr. Anchorage heading towards the conference room from the other direction with his trailing swarthy bodyguards. As I got close to him, I tripped and bumped into him. The guards quickly grabbed me and pushed me away with a glare. I made my apologies and stood out of the way, as they passed and then I followed into the press-room.

Mr. Anchorage read from a prepared statement and then took questions.

After a few softball questions from other reporters, I raised my hand.

"Mr. Anchorage, I appreciate the energy savings and reduction in carbon usages that will occur by a universal shift of MallMart to LED lights and your increase in organic products will also be a major boost to the organic industry that will create a more sustainable use of our environment. But considering your history of low wages for employees, lack of healthcare for your work force, your predatory and destructive practices with your suppliers that result in an imbalanced market that provides an unfair advantage to your stores at the cost to local stores, how are your current actions not just Green Washing?"

I could hear the murmur in the crowd of reporters as they expressed their shock at such a blunt and rude question. They could tell that this was not going to be a routine media event.

The CEO smiled with a false smile. I could see from the look in his eyes that he recognized me. "We provide more jobs than any other company across the US, which helps communities exist. Our providers are happy to distribute to us at the prices that we request and our business model allows people to buy what they need at cheap prices," he said.

"You are here announcing the benefits that you bring by increasing organics, but you are also the largest outlet for meat from Confined Animal Feeding Operations, which are bad for animal health, requiring a large amount of antibiotics, which is also damaging to human health. And this is not to mention the deterioration of the environment that results from these practices," I said.

"But people want cheap food," he fired back.

"Considering that you are one of the largest employers in the U.S., you could help solve this problem by paying your employees a living wage where they could afford meat from animals that are not kept in unhealthy conditions and damage the environment. You could be the solution to many of our sustainability issues by not being so greedy, providing a living wage, and sourcing all of your products from vendors that respect the environment that we depend upon," I said.

"That would damage our profits and destroy our business."

"I believe that you have quite a large margin to work with where you could be more equitable to your employees, competitors, and the environment. You don't have to dominate every market in the U.S. You could leave some market share for your competitors, and provide better working conditions for your employees. You could change the system to make it more equitable. The Mall family is already one of the richest in the world. What more do they need? They could take this opportunity to improve the quality of life for all workers in the U.S. and to source products in such a way as to leave a viable environment for future generations. But I wanted to ask a question on a different topic. Do you employ a security company called Black Cube and do you make a regular habit out of trying to kill journalists that push too hard to uncover your business practices as you did with me back in December?"

"That is an outrageous claim," he said vehemently. "We would never resort to bodily harm. We are a professional company, not the mafia."

I held up a slightly blurry picture printed on an 8 1/2 by 11 inch sheet of paper. "So why does that bodyguard next to you look exactly like this man that drove me off the road in Virginia?" It had turned out that the red truck coming up the mountain that came to my assistance also had a dash camera in his vehicle. It had caught a picture of the face of the driver of the vehicle that ran me off the road, and it was one of Mr. Anchorage's bodyguards this particular day.

"I think this session is at an end," he said as he quickly left the backside of the stage with his large bodyguards following him out the back door.

The media in the room swarmed over to me and started asking questions about my encounter in the mountains of Virginia, Black Cube, and what I had found out. I shared a bit of the more obvious information that they would be able to glean from the police report and some information about Black Cube. If they wanted to dig into that information it would provide some more pressure on MallMart and their security company, but I kept most of my information to myself for my own stories.

Then I went back to my hotel room to monitor the bug that I put into Mr. Anchorage's pocket when I ran into him.

Chapter 29

Threatened

St. Louis, Missouri
May 11

I returned to my hotel room and put on the headphones to listen to the recording device that digitally recorded everything that was picked up by the bug in Mr. Anchorage's pocket and also backed it up to the web through a wifi connection at the same time.

I had not been very secretive about where I was staying and my car was parked out front of the motel room. I had the window curtain open as I sat at the desk and played back the recording just after the press conference from about half an hour ago.

I could hear my own voice saying "…why does that bodyguard next to you look exactly like this man that drove me off the road in Virginia?"

And Mr. Anchorage say ,"I think this session is at an end," as there was a rustle of fabric as he hurried out the back door. You could hear the noise from the questions to me in the media room dropping away as he quickly walked down the hall to exit the building. I heard the sound of heavy door opening and then close behind them. I could just pick up the ambient noise that sounded like the echo of a parking garage that was just behind the conference rooms.

"How did he get that information about your company and the picture of you?" Mr. Anchorage demanded.

"I don't know. Maybe he's bluffing?" A deep eastern accented voice responded.

"I don't think that he's bluffing. He knows too much. You need to take him out this time and not make any mistakes. Get the job done!" he yelled.

"Yes, sir. I will take care of it." I could hear a car door closing and the squeal of tires as a car pulled out of the parking garage.

I continued to listen to the recording about 20 minutes delayed from real time as I watched out the window making notes of his further conversations after I made a phone call.

The CEO of MallMart apparently liked using the speakerphone, because I could hear both sides of the conversation just fine.

"Hey Mike, this is Roy Anchorage. We have a problem. This nosy reporter is looking into our operation. You better shut down that meat packing operation in Kansas for a while. Things are going to get a little messy and I want to keep everything quiet until this goes away."

"Are you taking care of the problem?" Mike asked.

"Yes, we have a team on that."

"How long will it take?"

"I would just shut the factory down for a couple of weeks. The problem should be gone today, but we want things to cool down in case there is an investigation before we pick operations back up."

"Will do. Let me know when we have the green light." Mike said as he hung up.

"Afternoon, George," Roy said to the phone.

"Hey Roy, glad you called. I wanted to give you an update on that price agreement with the packing industry."

"Actually George, I was calling about that. We are getting pressure from an investigative reporter. We should put a hold on that inside trading for a few weeks until this blows over," Roy said.

"Really, we just about have the regulators and the CEOs of the companies about to cave to our demands for that lower price even though it hurts their companies. They really want to deal with you and are scared that you will stop using their products."

"I'm not surprised. We have the power, but that deal is not going anywhere. Tell them that we want them to think about the offer for two weeks and then we'll call them back. Tell them not to contact us until we say we're ready. That should shut them up and also put the pressure on."

"OK, you're the boss. Are you sure the problem will blow over in a couple of weeks?"

"Yes, we should end the problem today, but we want to give it a little time until the smoke has cleared," Roy said as he hung up.

I was thinking that this problem they were ending was likely to be me.

It had been less than an hour since my confrontation with Roy Anderson when I saw a large black SUV drive up in the parking lot of the hotel. I felt nervous, but also satisfied that it was finally happening. I made sure that the audio was still recording Roy's continued conversations and hit record on the video cameras that I had set up around my hotel room and out on the balcony overlooking the parking lot.

I was staying in an old-style motel with outdoor entrances to three stories of rooms. Balconies wrapped around the building where you could walk around the floors to get to the various rooms and a couple of stairways ran down to the parking lot.

Four large swarthy men got out of the black SUV and headed straight to the stairs leading up to my room. I'm not sure how they knew where I was staying, but they seemed to be well-informed. I just watched as they came up to my third story room.

The first one kicked in my door and they quickly filed in the room taking multiple shots towards my image at the desk at my computer. There was the loud report of the bullets, the sound of shattering glass, and the smell of sulfur in the air.

Then, ten black-clothed people in full battle gear with SWAT on the back of their jackets came out of the woodwork. Six of them stepped out from around curtains and furniture in the room and four more emerged from the rooms on either side of mine with AK-47 aimed at the intruders. The new arrivals were shouting at the intruders to put up their hands and drop their guns. The four men looked surprised and dropped their handguns, being severely outnumbered, out armed, and outmaneuvered. They were pushed to the ground and their hands were secured behind their backs with zip ties as they were searched for other weapons. The vehicle outside was surrounded a few seconds later by police cars and more SWAT team members.

"All clear," one of them barked as the scene was secured.

I stepped out from my closet-like space behind the door and I found it hard to walk and not faint on my shaky legs.

A plain cloths police detective came in the door at that point and surveyed the scene. "Good job men. Greg, nice acting. You kept your cool," Detective O'Brien said.

"Thanks, do you mind if I sit down?" I said as I melted on to the couch.

"No problem. Do you recognize any of these men?" O'Brien asked gesturing to the four thugs now kneeling on the floor with their hands tied behind their backs.

"That one is the one that visited my house and threatened me. That one was the one that tried to drive me off the road based on the webcam photo. The third one I've seen around town a few places and the fourth escorted me from Anchorage's office in Kalamazoo," I said identifying all four of them.

"Great. It looks like we wrapped this one up tight. Their intent was definitely to kill you so we have solid evidence that they were executing a hit," he said.

"The mirror set-up and the protective closet with metal sheet reinforcement worked well. It looked like I was sitting at the desk with the mirror placed there while I was actually sitting next to the door in the secure box."

"They saw your image where they expected it when they kicked down the door and started firing. Now we have them all on attempted murder and combining that with your recordings, we have Anchorage on contracting a hit besides any of the other corporate issues you have dug up," O'Brien said. He checked his cell phone as I heard the buzz of text messages coming in.

"It looks like another team has picked up Anchorage and secured his computer," O'Brien said as he read through the texts. "And it looks like the third team has secured his headquarters in Kalamazoo, have detained the workers, and have started to confiscate their files and computers according to the warrant."

"It looks like we'll have a lot of data as this information comes to light. Thanks for the protection and all of your help with this case," I said.

"It'll take us a while to go through the evidence but we'll share our findings with John Hackett and you when we can after we have developed the criminal case on attempted murder and hiring a hit. Then you all can pursue the corporate malfeasance issues," he said as he left the room with the SWAT team leading the four thugs to a collection van.

Chapter 30

The Settlement

Indianapolis
September

It took about four months, but the State of Indiana brought attempted murder charges against the four thugs. Roy Anchorage was charged with conspiracy to commit murder. My recordings and the testimony from Detective O'Brien made this an open and shut case. Allowing it to play through to the point where the thugs thought they were firing on me made compelling video evidence in the courtroom, as did my recordings of Roy Anchorage giving the orders to end me.

The four thugs were tight lipped and were never willing to talk about anything dealing with the case. They each got life sentences with no probation and disappeared into the criminal justice system after that.

Roy Anchorage was willing to talk about MallMart's business practices to lighten his sentence. He was sentenced to 50 years in prison with a chance for parole. After that, he started to cooperate with John Hackett and me as we built our case against MallMart and the family that owned it. His testimony along with all of the information that was gleaned from the corporate computers made a very clear case against the company.

About six months after the criminal trial for attempted murder and conspiracy to commit murder, the final rulings came out about MallMart and its practices. The judge found that MallMart was guilty on multiple counts. The rulings came down one after another.

Guilty of corporate espionage

Guilty of price matching and a suite of other unfair business practices

Guilty of price manipulation

The judge fined the Mall brothers $2 billion dollars related to the charges of unfair business practices. They had to set up an endowment that would provide micro loans and moderate sized loans to "promote the health of small communities".

After the CEO of MallMart was handcuffed and taken off to jail, a lower manager named James Raydahl was promoted to CEO. He had been leading the charge for natural lighting, the conversion to CFLs, and the installation of solar panels at their stores. He was tasked by the courts, and therefore by the board of directors at MallMart, to make their stores more sustainable and to become part of the community. They changed their business practices. They continued to try to provide low prices, but not at the expense of the environment. They pushed for $15 minimum wage and universal healthcare in the United States. They created policies to push for 60% organic content in their grocery products and would only work now with vendors that supported sustainable agriculture and product development.

The MallMart business was founded on the ideal of sharing the profits and pride of the store with its employees. In their push for growth they had left that ideal behind and only focused on growth at all costs which created poor working conditions, destroyed small communities, and damaged the Earth. Now, because of the intervention of the courts, they were going back to their roots and providing for their workers and the communities where they operated.

Chapter 31

Going Local

Robertsville

April

Jenny and I were walking through Robertsville for First Friday pushing David in a stroller. The Downtown Association decided to celebrate their businesses by staying open late on the first Friday of each month. The shop owners had sales and the town hosted activities for the kids as well as music events around the square. It was festive and many people came in from the surrounding communities to join in this monthly celebration of our town.

I ran into James Caldwell, the town mayor while I was walking downtown to pick up fresh produce at the farmer's market. He was looking at a construction site in what used to be an old historic three-story building along the downtown square.

"Hey James, what're they building here?"

"They're renovating these old buildings as residential apartments. We were able to get a HUD redevelopment grant and partner with a private company to convert these old abandoned buildings into apartments."

"That's great! I had heard that one of the best practices for urban redevelopment is to build housing in the downtown area."

"Yeah. After we talked about how MallMart was draining our community, I attended an Urban Development conference. They were saying that downtown development can be energized by building apartments in the downtown area. It builds a shopper base within walking distance and it also improves safety downtown at night because there are more eyes on the streets. More people around and looking out their windows has been found to deter crime," he said.

"Great! Is all of this space going to be residential?"

"No. The bottom floors are reserved for commercial use. We already have a local baker and an Irish pub that have signed a lease on these two main floor areas."

"So we'll get a new bakery and an Irish Pub downtown. That's great! I would definitely frequent those places. I love fresh baked bread and an Irish pub is a nice setting to sit and work or just hang out. Are they going to have food as well?" I asked.

"Yes, so it'll also be a restaurant downtown. We just have one commercial space left to lease. The construction should be done in about two months."

We continued on down the road past a couple of small shops.

"George and Nancy, how're you?" I said. The old grocery store couple and their two grown kids had a table outside of their old store.

"We're doing well. Our food coop idea is moving forward. We have 400 member/owners already, which is enough get the doors open. We're running a capital campaign to collect loans from our member owners where they can invest in their local grocery and their community. We received a community development loan from MallMart that will also help to get the store open," Joanne, Nancy's daughter said.

"We hope to complete the capital campaign in the coming month. We'll have to get a loan for the rest of the funds that we need from our local bank, but we should be able to start renovations in the next couple of months. With the member equity that we have accumulated, we hired consultants to develop a store design and architecture plans for the renovations. Look at these designs," Steven said as he pointed to an easel displaying realistic looking drawings of the exterior and interior of the store.

"Those look great. This is a small modern store. It would be wonderful to shop in there!" I said.

"We also had a consultant draw up an interior design for the grocery store. There is a science to how you should put products in a grocery store. This will modernize the store and make it flow better for our customers. We're looking forward to the renovations and hope to have our grand opening next summer," Steven said.

"That's wonderful! I can't believe that we will get our grocery story back and it will be even better than before! Sign us up as members," Jenny said.

"It's nice to have our town back and it feels like it's growing even better!" I said.

The End

Glossary

Brownfield - A contaminated property that usually had some type of industrial process on it that lead to the contamination. Today, these areas are often abandoned and no one wants to purchase them or develop them because they may inherit the cost of previous environmental problems.

Car Share Program – This program is usually run through a car rental agency, where cars are parked on the businesses property (like a university campus) and the members of that business (faculty, staff, and students) have access to the car. They can rent the car for an hour or up to days. This makes driving transportation more accessible. It also allows younger people to participate in car rental because the insurance covers everything affiliated with the University.

Cooperative Markets – A different business model where member/owners buy into a store. They then get voting rights to determine the direction of the store. Cooperative markets in the 1970s and 1980s were mostly volunteer driven while today they are a small business with full-time employees which is important for keeping the store open.

Economies of Scale – When making products in bulk or buying them in bulk reduces the price of the individual item. There is an efficiency increase to making large numbers of the same item that causes the price of that item to decrease.

Ecotourism – Tourism that is focused on enjoyment of the natural setting. This helps to provide work and income for local people while enabling visitors to enjoy the beautiful scenery. We have to be careful not to overwhelm these natural habitats that everyone likes to visit.

Environmental Justice – Making sure that all people of any race, ethnicity, gender, or religion have the same access to goods and services in the system without any stigma. An example of environmental injustice is that most landfills have been located near high minority populations. Groups are trying to change this occurrence so that we are more just in the United States relative to pollution burden.

Externalities – A cost or benefit in an economic system that is not calculated into the cost of the item. This is often a harmful event, like pollution or soil degradation, which does not get counted into the price of an item.

Green Washing – This occurs when a company sells its products as environmentally helpful when in reality there is no benefit for the environment and sometimes the products even cause harm to the environment. Many corporations recognize that the public is interested in living lighter on the Earth and may take advantage of that concern by marketing their products to that demographic even though their products are not better for the environment.

Gross National Happiness – Bhutan developed this index to measure the well-being and contentment of their populace in 1972. Since its development, many other countries have been looking to happiness as a meaningful metric.

LEED Certification –Leadership in Energy and Environmental Design. A Green Building Council certification program for new construction and renovation. It has been successful in making sustainable best practices part of the mainstream in construction design.

Living Building Challenge – Beyond LEED Certification in trying to design and build buildings that give back to the environment. They produce more energy than they use and they have systems in place to naturally clean water and run-off.

Natural Step – Developed in 1989, this is a science-based approach to sustainability developed by Swedish Professor Karl-Henrik Robèrt that focuses on reducing unnatural concentrations of substances like CO_2 and heavy metals, reducing substances produced by society like antibiotics, plastics, and endocrine disruptors, reducing the degradation of the physical environment, and removing structural barriers to health, influence, competence, impartiality, and meaning.

Permaculture – A set of design principles that researchers have learned from nature that involves systems thinking to grow edible food in high concentrations in a sustainable fashion. These plantings usually involve perennial plants that produce good quality food for humans, and when it is done correctly, it provide more benefit as it matures. David Holmgren and Bill Mollison at the University of Tasmania developed this concept in 1978.

Quality of Life – A subjective term where individuals define their level of wellbeing. Many factors including emotional, physical, and material wellbeing going into this concept and the important factors differ for each person. It is becoming an important metric for the quality of a place and a guide for how to improve living conditions through a locally driven process.

Resource Conservation and Recovery Act (RCRA) – A law that was enacted in 1976 replacing the Solid Waste Disposal Act of 1965, which congress enacted to deal with municipal and industrial waste. It was meant to protect human health and the natural environment, conserve energy and natural resources, reduce waste at its source and through recycling, and manage waste in an environmentally safe way.

Superfund – A fund that was developed from a tax on polluting companies that is used to clean up past pollution events. Superfund sites are usually the largest and most severely contaminated sites in the United States.

Sustainable Development Goals – A set of 17 goals developed by the United National General Assembly in 2015 that focus on social, economic, and environmental development issues. These goals focus on improving such things as poverty, hunger, education, sanitation, and inequality around the world for benefit to all people.

Sustainable Forestry/Agriculture –Forestry or agriculture practices that don't deplete the soil that is necessary for their growth. These practices protect the functioning ecosystem that supports the growth of trees or agricultural plants.

Upcycle – Taking recycled material and making a product that is more valuable than the original material. Often times, this is taking recycled material as a raw material and making artwork from it. This provides value to our waste streams, keeping them out of the landfill while providing economic value.

Wind Turbine –In the past, windmills were use to grind corn or pump water. Today, the wind turbine has a coil of copper wire placed inside of magnets so that when the wind blows the coil of wire rotates causing electrons to flow. This creates electricity that passes down the line.

References

EndNotes Chapter 1: The Demise of Small Towns

[1] Speer, J.H. 2018. Exposé on Climate Change. Kendall Hunt Publishing Company. 291pp.

[2] Stone, K.E., 1997. Impact of the Wal-Mart phenomenon on rural communities. Increasing understanding of public problems and policies, 1997, pp.1-21.

[3] Miczek, J.E. 2016. Small towns devastated after Wal-Mart Stores Inc decimates mom-and-pop shops, then packs up and leaves: "They ruined our lives". Bloomberg News. https://business.financialpost.com/news/retail-marketing/small-towns-devastated-after-wal-mart-stores-inc-decimates-mom-and-pop-shops-then-packs-up-and-leaves-they-ruined-our-lives. Downloaded 1/21/2019.

EndNotes Chapter 2: The Homestead

[1] K-State Research and Extension News. 2018. As corn and wheat prices drop, farmers look to alternative crops. https://www.ksre.k-state.edu/news/stories/2018/11/alternative-crops.html. Downloaded 1/21/2019.

[2] Doering, C. 2016. Iowa farmers planting fruits, vegetables over corn, soybeans. Des Moines Register. https://www.desmoinesregister.com/story/money/agriculture/2016/08/19/iowa-farmers-planting-fruits-vegetables-over-corn-soybeans/88569432/. Downloaded 1/21/2019.

[3] EPI. 2019. The State of Working America. http://stateofworkingamerica.org/chart/swa-wages-table-4-43-ceo-compensation-ceo/. Downloaded 1/19/2019.

[4] Ingraham, C. 2017. The richest 1 percent now owns more of the country's wealth than at any time in the past 50 years. https://www.washingtonpost.com/news/wonk/wp/2017/12/06/the-richest-1-percent-now-owns-more-of-the-countrys-wealth-than-at-any-time-in-the-past-50-years/?noredirect=on&utm_term=.86597a6c2a47. Washington Post.

[5] Wolff, E.N., 2017. Household Wealth Trends in the United States, 1962 to 2016: Has Middle Class Wealth Recovered? (No. w24085). National Bureau of Economic Research.

EndNotes Chapter 3: Making the Pitch

[1]World Wildlife Fund. 2019. Urban Solutions for a Living Planet. http://wwf.panda.org/our_work/projects/one_planet_cities/urban_solutions/. Downloaded 1/21/2019.

EndNotes Chapter 5: The Sustainability Tour

[1] Watson, S.K. 2018. China Has Refused To Recycle The West's Plastics. What Now? https://www.npr.org/sections/goatsandsoda/2018/06/28/623972937/china-has-refused-to-recycle-the-wests-plastics-what-now. Downloaded 1/19/2019.

[2] WTO. 2017. China's import ban on solid waste queried at import licensing meeting. https://www.wto.org/english/news_e/news17_e/impl_03oct 17_e.htm. Downloaded 1/19/2019.

[3] Brooks, A.L., Wang, S. and Jambeck, J.R., 2018. The Chinese import ban and its impact on global plastic waste trade. Science Advances, 4(6), p.eaat0131.

[4] Latimer, J.C., Van Halen, D., LA, S.K., Weaver, P. and Foxx, H., 2016. Soil lead testing at a high spatial resolution in an urban community garden: a case study in relic lead in Terre Haute, Indiana. Journal of environmental health, 79(3), p.28.

EndNotes Chapter 6: Peak Oil

[1] Hopkins, R., 2008. *The transition handbook. From oil dependency to local resilience.* Green Books Ltd, Foxhole, Dartington, Totnes, Devon TQ9 6EB.

[2] Kinsale, Ireland. 2019. http://www.bluehavenkinsale.com/. Downloaded 2/23/2019.

[3] Woodham-Smith, C. 1962. The Great Hunger: Ireland 1845-1849. London: Hamish Hamilton. 510 pages.

[4] Wikipedia. 2017. Hubbert's Peak Theory. https://en.wikipedia.org/wiki/Hubbert_peak_theory Downloaded 11/15/2017.

[5] McCarthy, N. 2017. Solar Employs More People In U.S. Electricity Generation Than Oil, Coal And Gas Combined. Forbes Magazine. Jan 25, 2017, 08:30am. https://www.forbes.com/sites/niallmccarthy/2017/01/25/u-

s-solar-energy-employs-more-people-than-oil-coal-and-
gas-combined-infographic/#73ea9ac62800

[6] DOE 2017. US Energy and Employment Report.
https://www.energy.gov/sites/prod/files/2017/01/f34/2017
%20US%20Energy%20and%20Jobs%20Report_0.pdf

[7] Jackson, R.B., Vengosh, A., Darrah, T.H., Warner, N.R.,
Down, A., Poreda, R.J., Osborn, S.G., Zhao, K. and Karr,
J.D., 2013. Increased stray gas abundance in a subset of
drinking water wells near Marcellus shale gas extraction.
Proceedings of the National Academy of Sciences, 110(28),
pp.11250-11255.

Endnotes for Chapter 7: The Healthcare System

[1] Department for Professional Employees Research Department.
2016. The U.S. Health Care System: An International Perspective. Fact Sheet 2016. https://dpeaflcio.org/programs-
publications/issue-fact-sheets/the-u-s-health-care-system-
an-international-perspective/ Downloaded 3/3/2019.

[2] Ortiz-Ospina, Esteban and Roser, Max. 2019. "Financing
Healthcare". Published online at OurWorldInData.org. Retrieved from: 'https://ourworldindata.org/financing-
healthcare' [Online Resource]. Downloaded 3/20/2019.

EndNotes Chapter 9: Into the Mouth of the Beast

[1] Quinn. 2016. 5 Best and Worst Jobs at Walmart.
https://www.huffingtonpost.com/gobankingrates/5-best-
and-worst-jobs-at_b_9202602.html. Downloaded
11/17/2017.

EndNotes Chapter 11: Wealth Inequality

[3] uTrend.tv. 2017. 9 out of 10 Americans Are Completely Wrong about This Mind Blowing Fact. http://utrend.tv/v/9-out-of-10-americans-are-completely-wrong-about-this-mind-blowing-fact/ Downloaded 11-15-2017.

[4] Norton, M.I. and Ariely, D., 2011. Building a better America—One wealth quintile at a time. Perspectives on psychological science, 6(1), pp.9-12.

[5] Keister, L.A. and Moller, S., 2000. Wealth inequality in the United States. Annual Review of Sociology, 26(1), pp.63-81.

EndNotes Chapter 12: Lap of Luxury

[4] Peterson-Withorn, C. 2016. Forbes 400: The Full List Of The Richest People In America 2016. https://www.forbes.com/sites/chasewithorn/2016/10/04/forbes-400-the-full-list-of-the-richest-people-in-america-2016/#6034ae8922f4. Downloaded 11/17/2017.

[5] Encyclopedia of Nations. 2017. Dominican Republic - Poverty and wealth. http://www.nationsencyclopedia.com/economies/Americas/Dominican-Republic-POVERTY-AND-WEALTH.html Downloaded November 19th, 2017.

[6] Lawlor, O.A. 2016. Taíno: Indigenous Caribbeans.http://www.blackhistorymonth.org.uk/article/section/pre-colonial-history/taino-indigenous-caribbeans/ B:M@30. Downloaded November 20th, 2017.

[7] Livi-Bacci, M. 2006. "The Depopulation of Hispanic America after the Conquest". Population and Development Review.

32 (2): 208–213. ISSN 0098-7921. JSTOR 20058872. Retrieved 2017-03-02.

[8] Alchon, S.A. 2003. A pest in the land: New world epidemics in a global perspective. Albuquerque: University of New Mexico Press. pp. 164–167. *ISBN 978-0-8263-2870-0*.

EndNotes Chapter 13: The Mountain

[9] Speer, J.H., Orvis, K.H., Grissino-Mayer, H.D., Kennedy, L.M. and Horn, S.P., 2004. Assessing the dendrochronological potential of Pinus occidentalis Swartz in the Cordillera Central of the Dominican Republic. *The Holocene*, *14*(4), pp.563-569.

EndNotes Chapter 14: Walkable Bikeable Cities

[1] EPA. 2017. Office of Land and Emergency Management (OLEM) Accomplishment Reports and Benefits Google Archive Snapshot Jan 19, 2017. https://19january2017snapshot.epa.gov/aboutepa/office-land-and-emergency-management-olem-accomplishment-reports-and-benefits_.html Downloaded November 18th, 2017.

[2] Buettner, D. 2017. The happiest cities in America, according to new research. http://www.cnn.com/2017/11/22/health/happiest-cities-blue-zones/index.html Downloaded November 22nd, 2017.

[3] WalkScore. 2017. https://www.walkscore.com/ Downloaded November 18th, 2017.

EndNotes Chapter 15: The Wedding.

[1] Meme Police. 2017. Misleading With Inflation Since 1978. http://memepoliceman.com/misleading-with-inflation-since-1978/ Downloaded 3/20/2019.

EndNotes Chapter 16: The Hoosier Environmental Council

[1] McCarthy, N., 2017. Solar Employs More People In U.S. Electricity Generation Than Oil, Coal And Gas Combined. https://www.forbes.com/sites/niallmccarthy/2017/01/25/u-s-solar-energy-employs-more-people-than-oil-coal-and-gas-combined-infographic/#286a4be72800 Forbes Magazine. Downloaded November 19th, 2017.

[2] Hoosier Environmental Council. https://www.hecweb.org/ Downloaded November 19th, 2017.

[3] Hirtenstein, A. 2016. Clean-Energy Jobs Surpass Oil Drilling for First Time in U.S. https://www.bloomberg.com/news/articles/2016-05-25/clean-energy-jobs-surpass-oil-drilling-for-first-time-in-u-s Bloomberg Press. Downloaded November 19th, 2017.

Endnotes Chapter 17: AASHE Conference

[1] AASHE. 2018. Association for the Advancement of Sustainability in Higher Education. http://www.aashe.org/. Downloaded October 27th, 2018.

[2] MAX. 2018. Metropolitan Area Express. https://trimet.org/max/. Downloaded October 27th, 2018.

[3] PLAN. 2018. Post Landfill Action Network.
https://www.postlandfill.org/. Downloaded October 27th, 2018.

[4] Willamette University Green Revolving Fund. 2018.
https://willamette.edu/about/sustainability/green-fund/index.html. Downloaded October 27th, 2018.

[5] The Natural Step. 2018. https://thenaturalstep.org/. Downloaded October 27th, 2018.

[6] United Nations Millennium Development Goals. 2018.
http://www.un.org/millenniumgoals/. Downloaded October 27th, 2018.

[7] UN Sustainable Development Goals.
https://www.un.org/sustainabledevelopment/climate-change-2/. Downloaded October 27th, 2018.

Endnotes Chapter 20: Cradle to Cradle

[1] McDonough, W. and Braungart, M., 2002. *Cradle to cradle: Remaking the way we make things*. North Point Press; 1st edition (April 22, 2002). 193pp.

[2] McDonough, W. and Braungart, M., 2013. *The upcycle: Beyond sustainability--designing for abundance*. Macmillan. 227pp.

EndNotes Chapter 22: Rephrasing Climate Change Discussions

[1] Stoknes, P.E. 2017. How to transform apocalypse fatigue into action on global warming.
https://www.ted.com/talks/per_espen_stoknes_how_to_transform_apocalypse_fatigue_into_action_on_global_warming Downloaded November 19th, 2017.

[2] Mann, M.E., Bradley, R.S. and Hughes, M.K., 1999. Northern hemisphere temperatures during the past millennium: Inferences, uncertainties, and limitations. *Geophysical research letters*, *26*(6), pp.759-762.

[3] Wilson, R., Anchukaitis, K., Briffa, K.R., Büntgen, U., Cook, E., D'arrigo, R., Davi, N., Esper, J., Frank, D., Gunnarson, B. and Hegerl, G., 2016. Last millennium northern hemisphere summer temperatures from tree rings: Part I: The long term context. *Quaternary Science Reviews*, *134*, pp.1-18.

[4] Anchukaitis, K.J., Wilson, R., Briffa, K.R., Büntgen, U., Cook, E.R., D'Arrigo, R., Davi, N., Esper, J., Frank, D., Gunnarson, B.E. and Hegerl, G., 2017. Last millennium Northern Hemisphere summer temperatures from tree rings: Part II, spatially resolved reconstructions. *Quaternary Science Reviews*, *163*, pp.1-22.

EndNotes Chapter 24: Downhome

[1] Our Children's Trust. 2018. Juliana v. U.S. – Climate Lawsuit.
https://www.ourchildrenstrust.org/us/federal-lawsuit/.
Downloaded November 1st, 2018.

[2] National Association of Insurance Commissioners. 2008. The Potential Impact of Climate Change on Insurance Regulation. http://www.naic.org/documents/cipr_potential_impact_climate_change.pdf. Retrieved 2-23-2018.

[3] National Association of Insurance Commissioners. 2019. Climate Change and Risk Disclosure. https://www.naic.org/cipr_topics/topic_climate_risk_disclosure.htm. Downloaded 3/20/2019.

EndNotes Chapter 25: Legacy

[1] Poyourow, Joanne. 2005. Legacy: A story of hope from a time of environmental crisis. Virtualbookworm.com Publishing Inc. 383pp.

EndNotes Chapter 26: The Very Air We Breath

[1] Pigg, J. and Speer, J.H. 2015. Unpublished Data. Pollution Loads of Heavy Metals and Resilience in the Wabash River Ecosystem. Poster Presented at the Summer Undergraduate Research Experience in August 2015.

[2] EPA. 2013. Aquatic life ambient water quality criteria for ammonia – Freshwater. EPA 822-R-13-001. https://www.epa.gov/sites/production/files/2015-08/documents/aquatic-life-ambient-water-quality-criteria-for-ammonia-freshwater-2013.pdf. 255pp.

3 Henry, D.W. ND. The Story of Fort Harrison. A paper presented to the Fort Harrison Country Club.

Index

D

Q

R

Y

Z

ABOUT THE AUTHOR

 Dr. James H. Speer is a Professor of Geography and Geology at Indiana State University. He received his bachelors and master's degree from the University of Arizona in Geosciences and his PhD from the University of Tennessee in Geography. He is a biogeography who uses tree-ring to reconstruct environmental variables such as fire history and insect outbreaks. Through his years of studying environmental history he has realized that humans are operating outside of the natural range of variability for most natural systems which has motivated him to give back to society by being a champion for sustainability at Indiana State University and in the Wabash Valley. Dr. Speer is a Senior Scholar for the Institute for Community Sustainability, which was established in February 2012. He is the President for Our Green Valley Alliance for Sustainability, on the board for the Terre Foods Cooperative Market, on the Tree Advisory Board for ISU, is a past president of the Geography Educator's Network of Indiana, and is a past president of the Tree-Ring Society. He lives in Terre Haute Indiana with his wife who also teaches at ISU with a PhD in Anthropology and their two sons Leif and Lewis.